Handbook for Ironmongers

A Glossary

Of

Ferrous Metallurgy Terms

Hand Tools in History Series

- Volume 6: Steel- and Toolmaking Strategies and Techniques before 1870
- Volume 7: Art of the Edge Tool: The Ferrous Metallurgy of New England Shipsmiths and Toolmakers 1607 to 1882
- Volume 8: The Classic Period of American Toolmaking, 1827-1930
- Volume 9: An Archaeology of Tools: The Tool Collections of the Davistown Museum
- Volume 10: Registry of Maine Toolmakers
- Volume 11: Handbook for Ironmongers: A Glossary of Ferrous Metallurgy Terms: A Voyage through the Labyrinth of Steel- and Toolmaking Strategies and Techniques 2000 BC to 1950
- Volume 13: Tools Teach: An Iconography of American Hand Tools

Handbook for Ironmongers

A Glossary

of

Ferrous Metallurgy Terms

A Voyage through the Labyrinth of Steel- and Toolmaking Strategies and Techniques 2000 BC to 2000 AD

and the Cascading Industrial Revolutions they Engendered

H. G. Brack, Editor

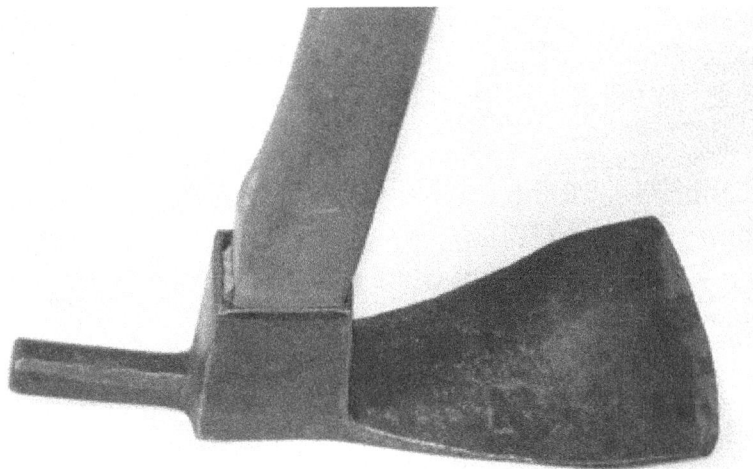

Davistown Museum
Museum Publication Series Volume 11

© Davistown Museum 2013
ISBN 978-0-9892678-0-9

Davistown Museum

Second Edition

Front cover illustrations: Clockwise starting at the top left

Small hand adz from the collection of the Davistown Museum ID: TBW1010. Forged iron and wood, c. 1740 – 1780, 4 ¼" long with a 4 ¾" handle and 1" wide blade.

Basket-making knife from the collection of the Davistown Museum ID: 011006T1. Blister steel, circa 1700, 6 7/8" long and ½" wide. This is a rare variation of the crooked knife and was solely used for splitting the wood for baskets of ash or other woods.

Grub hoe from the collection of the Davistown Museum ID: TAB1011. Forged iron, 17th or 18th century, 4" wide and 6" high.

Fishing spear from the collection of the Davistown Museum ID: TAB1015A. Forged iron, mid-17th century or earlier, 6 ½" long. Probably a French trade item.

Belt hatchet from the collection of the Davistown Museum ID: 111206T1. Iron and weld steel with a wood handle, c. 1680-1750, 8" long head, 2 ¼" wide cutting edge, 16 ¼" long handle. A tomahawk with traditional weld steel forging and a more recent handle.

Frontispiece illustration: Peen adz from the collection of the Davistown Museum ID: 020807T1. 10" long, 4 1/8" wide adz head, 2 ¼" long and ¾" diameter beveled peen, and 28" long handle. Signed by J. F. Ames, who made edge tools in Richmond, Maine, c. 1855. He is listed in the 1855 Maine Business Directory as an edge toolmaker.

Cover design by Cory R. Courtois, Pages Plus (www.pagesplusme.com)

This publication was made possible by a donation from Barker Steel LLC.

Pennywheel Press
P.O. Box 144
Hulls Cove, ME 04644

Preface

Davistown Museum *Hand Tools in History*

One of the primary missions of the Davistown Museum is the recovery, preservation, interpretation, and display of the hand tools of the maritime culture of Maine and New England (1607-1900). The *Hand Tools in History* series, sponsored by the museum's Center for the Study of Early Tools, plays a vital role in achieving the museum mission by documenting and interpreting the history, science, and art of toolmaking. The Davistown Museum combines the *Hand Tools in History* publication series, its exhibition of hand tools, and bibliographic, library, and website resources to construct an historical overview of steel- and toolmaking strategies and techniques used by the edge toolmakers of New England's wooden age. Included in this overview are the roots of these strategies and techniques in the early Iron Age, their relationship with modern steelmaking technologies, and their culmination in the florescence of American hand tool manufacturing in the last half of the 19th century.

Background

During over 40 years of searching for New England's old woodworking tools for his Jonesport Wood Company stores, curator and series author H. G. Skip Brack collected a wide variety of different tool forms with numerous variations in metallurgical composition, many signed by their makers. The recurrent discovery of forge welded tools made in the 18th and 19th centuries provided the impetus for founding the museum and then researching and writing the *Hand Tools in History* publications. In studying the tools in the museum collection, Brack found that, in many cases, the tools seemed to contradict the popularly held belief that all shipwrights' tools and other edge tools used before the Civil War originated from Sheffield and other English tool-producing centers. In many cases, the tools that he recovered from New England tool chests and collections dating from before 1860 appeared to be American-made rather than imported from English tool-producing centers. Brack's observations and the questions that arose from them led him to research the topic and then to share his findings in the *Hand Tools in History* series.

Hand Tools in History Publications

- Volume 6: *Steel- and Toolmaking Strategies and Techniques before 1870* explores ancient and early modern steel- and toolmaking strategies and techniques, including those of early Iron Age, Roman, medieval, and Renaissance metallurgists and toolmakers. Also reviewed are the technological innovations of the Industrial Revolution, the contributions of the English industrial revolutionaries to the evolution of the factory system of mass production with interchangeable parts, and the development of bulk steelmaking processes and alloy steel technologies in the latter half of the 19th century. Many of these technologies play a role in the florescence of American ironmongers and toolmakers in the 18th and 19th centuries. Author H. G. Skip Brack cites archaeometallurgists such as Barraclough, Tylecote, Tweedle, Smith, Wertime, Wayman, and many others as useful guides for a journey through the pyrotechnics of ancient and modern metallurgy. Volume 6 includes an extensive bibliography of resources pertaining to steel- and toolmaking techniques from the early Bronze Age to the beginning of bulk-processed steel production after 1870.

- Volume 7: *Art of the Edge Tool: The Ferrous Metallurgy of New England Shipsmiths and Toolmakers* explores the evolution of tool- and steelmaking techniques by New England's shipsmiths and edge toolmakers from 1607-1882. This volume uses the construction of Maine's first ship, the pinnace *Virginia*, at Fort St. George on the Kennebec River in Maine (1607-1608), as the iconic beginning of a critically important component of colonial and early American history. While there were hundreds of small shallops and pinnaces built in North and South America by French, English, Spanish, and other explorers before 1607, the construction of the *Virginia* symbolizes the very beginning of New England's three centuries of wooden shipbuilding. This volume explores the links between the construction of the *Virginia* and the later flowering of the colonial iron industry; the relationship of 17th, 18th, and 19th century edge toolmaking techniques to the steelmaking strategies of the Renaissance; and the roots of America's indigenous iron industry in the bog iron deposits of southeastern Massachusetts and the many forges and furnaces that were built there in the early colonial period. It explores and explains this milieu, which forms the context for the productivity of New England's many shipsmiths and edge toolmakers, including the final flowering of shipbuilding in Maine in the 19th century. Also included is a bibliography of sources cited in the text.

- Volume 8: *The Classic Period of American Toolmaking 1827-1930* considers the wide variety of toolmaking industries that arose after the colonial period and its robust tradition of edge toolmaking. It discusses the origins of the florescence of American toolmaking not only in English and continental traditions, which produced gorgeous hand tools in the 18th and 19th centuries, but also in the poorly documented and often unacknowledged work of New England shipsmiths, blacksmiths, and toolmakers. This volume explicates the success of the innovative American factory system, illustrated by an ever-expanding repertoire of iron- and steelmaking strategies and the widening variety of tools produced by this factory system. It traces the vigorous growth of an

American hand toolmaking industry that was based on a rapidly expanding economy, the rich natural resources of North America, and continuous westward expansion until the late 19th century. It also includes a company by company synopsis of America's most important edge toolmakers working before 1900, an extensive bibliography of sources that deal with the Industrial Revolution in America, special topic bibliographies on a variety of trades, and a time line of the most important developments in this toolmaking florescence.

- Volume 9: *An Archaeology of Tools* contains the ever-expanding list of tools in the Davistown Museum collection, which includes important tools from many sources. The tools in the museum exhibition and school loan program that are listed in Volume 9 serve as a primary resource for information about the diversity of tool- and steelmaking strategies and techniques and the locations of manufacturers of the tools used by American artisans from the colonial period until the late 19th century.

- Volume 10: *Registry of Maine Toolmakers* fulfills an important part of the mission of the Center for the Study of Early Tools, i.e. the documentation of the Maine toolmakers and planemakers working in Maine. It includes an introductory essay on the history and social context of toolmaking in Maine; an annotated list of Maine toolmakers; a bibliography of sources of information on Maine toolmakers; and appendices on shipbuilding in Maine, the metallurgy of edge tools in the museum collection, woodworking tools of the 17th and 18th centuries, and a listing of important New England and Canadian edge toolmakers working outside of Maine. This registry is available on the Davistown Museum website and can be accessed by those wishing to research the history of Maine tools in their possession. The author greatly appreciates receiving information about as yet undocumented Maine toolmakers working before 1900.

- Volume 11: *Handbook for Ironmongers: A Glossary of Ferrous Metallurgy Terms* provides definitions pertinent to the survey of the history of ferrous metallurgy in the preceding five volumes of the *Hand Tools in History* series. The glossary defines terminology relevant to the origins and history of ferrous metallurgy, ranging from ancient metallurgical techniques to the later developments in iron and steel production in America. It also contains definitions of modern steelmaking techniques and recent research on topics such as powdered metallurgy, high resolution electron microscopy, and superplasticity. It also defines terms pertaining to the growth and uncontrolled emissions of a pyrotechnic society that manufactured the hand tools that built the machines that now produce biomass-derived consumer products and their toxic chemical byproducts. It is followed by relevant appendices, a bibliography listing sources used to compile this glossary, and a general bibliography on metallurgy. The author also acknowledges and discusses issues of language and the interpretation of terminology used by ironworkers over a period of centuries. A compilation of the many definitions related to iron and steel and their changing meanings is an important component of our survey of the history of the steel- and toolmaking strategies and techniques and the relationship of these traditions to the accomplishments of New England shipsmiths and their offspring, the edge toolmakers who made shipbuilding tools.

- Volume 13 in the *Hand Tools in History* series explores the iconography (imagery) of early American hand tools as they evolve into the Industrial Revolution's increased diversity of tool forms. The hand tools illustrated in this volume were selected from the Davistown Museum collection, most of which are cataloged in *An Archaeology of Tools* (Volume 9 in *Hand Tools in History*), and from those acquired and often sold by Liberty Tool Company and affiliated stores, collected during 40+ years of "tool picking." Also included are important tools from the private collections of Liberty Tool Company customers and Davistown Museum supporters. Beginning with tools as simple machines, reviews are provided of the metallurgy and tools used by the multitasking blacksmith, shipsmith, and other early American artisans of the Wooden Age. The development of machine-made tools and the wide variety of tool forms that characterize the American factory system of tool production are also explored. The text includes over 800 photographs and illustrations and an appendix of the tool forms depicted in Diderot's *Encyclopedia*. This survey provides a guide to the hand tools and trades that played a key role in America's industrial renaissance. The iconography of American hand tools narrates the story of a cascading series of Industrial Revolutions that culminate in the Age of Information Technology.

The *Hand Tools in History* series is an ongoing project; new information, citations, and definitions are constantly being added as they are discovered or brought to the author's attention. These updates are posted weekly on the museum website and will appear in future editions. All volumes in the *Hand Tools in History* series are available as bound soft cover editions for sale at the Davistown Museum, Liberty Tool Co., local bookstores and museums, or by order from www.davistownmuseum.org/publications.html, Amazon.com, Amazon.co.uk, CreateSpace.com, Abebooks.com, and Albris.com.

Table of Contents

Introduction

Handbook for Ironmongers: A Glossary of Ferrous Metallurgy Terms is published by The Davistown Museum as part of its *Hand Tools in History* publication series. This glossary is an outgrowth of the research done during the investigation of the metallurgy and metallography of many of the earlier (pre-1860) tools in the museum collection. The many questions raised by the survival of these tools led to the review and reevaluation of the art of edge toolmaking and the steel- and toolmaking strategies and techniques upon which it was based. Also prompting the publication of this glossary as a separate component of the museum publication series is the contemporary interest in the revival of older techniques of making iron, steel, and hand tools and the relevance of this information at the dawn of the era of the creative economy. The urgent need to maintain or reestablish sustainable environments and lifestyles in the coming age of biocatastrophe provides additional impetus for the compilation of this glossary.

Language & Interpretation

Many of the definitions and descriptions in this glossary and other *Hand Tools in History* publications involve attempts to recreate, retrieve, and uncover older meanings, many of which pertain to technologies that no longer exist. The editor's journey through the labyrinths of ancient metallurgical traditions, especially edge toolmaking, resulted in the compilation of this glossary, which combines modern scientific knowledge about ferrous metallurgy with necessarily subjective interpretations of what artisans and edge toolmakers thought and knew centuries ago, when rule of thumb and intuition dominated steel- and toolmaking strategies and techniques. Early traditions and techniques are defined by using a combination of the concepts and terminology of modern ferrous metallurgy, the writings of contemporary archaeometallurgists, and the fragmentary descriptions of the past to explain the metallurgical constituents of tools that survive from early times without any written records to describe how they were made. In compiling this glossary and attempting to explain early steel- and toolmaking strategies and techniques, the editor encountered the following concerns that impact our current understanding and interpretation of ferrous metallurgy terms.

Early Techniques

Before the evolution of modern science and the consequent development of an analytical understanding of the chemistry of ferrous metallurgy and the microstructures of iron and steel (metallography), the knowledge of successful steel- and toolmaking strategies and techniques was based on the repetitive, intuitive, empirical methodologies of smelting, carburizing (or decarburizing,) quenching, tempering, and annealing. These methodologies often produced iron and steel, including tools, of a surprisingly high quality. Early techniques were not based on theoretical or scientific conceptions but on trial and error methods and are frequently referred to as "rule of thumb" techniques (Brearley 1933, Barraclough 1984a). Intuition, ritual, and generations of empirical experience played an important role in the art of the ironmonger, hence our inclusion of "kan of the ironmonger," a term which describes the kind of knowledge necessary for the successful steel- and toolmaking that predates the era of bulk steel production (>1870). A Japanese word and concept, the swordsmith's "kan" was a combination of ritual and empirical experience, a methodology of swordsmithing that was beyond words or written language. The primary evidence for the technical finesse of early toolmakers is the tools they made, which are the legacy of their uncanny ability to forge hot iron into steeled tools.

Secrecy

Before the age of modern science, intuitive and rule of thumb knowledge of steelmaking techniques involved implementation of steelmaking strategies that were often unarticulated, closely guarded secrets manifested in fiery rituals of forging that sometimes verged on magic. Metallurgists, such as sword cutters and knife-makers, edge toolmakers, and forge masters who specialized in raw steel production, did not share the knowledge of their craft traditions in an open society characterized by free exchange of information. An important objective of any community or culture with an ability to make superior weapons, tools, iron, or steel would have been to prevent competing cultures and artisans from acquiring their skills. Until the 19th century, few attempted to systemize and record knowledge of steel- and toolmaking strategies and techniques. When such writing did begin to appear (Biringuccio [1540] 1990), it was of little use to practicing blacksmiths, swordsmiths, or edge toolmakers, who operated in a closed society that was based on mystery, secrecy, and conjuration, i.e. the production of magical effects (steeling) by natural empirical means (the carburizing or decarburizing of iron).

The Changing Meaning of Technical Vocabulary

The meaning of words pertaining to the history of steel- and toolmaking strategies and techniques varies from one cultural community to another and also changes over time. Even today, the terms "German steel," "wrought iron," "malleable iron," and "low carbon steel" have a variety of meanings no longer consistent with earlier terminology. In many cases, contemporary writers differ as to the meaning of these words, often not even attempting to define a generic term such as "malleable iron," which, it can be argued, has no specific meaning at all. Many of the terms used in this glossary are no longer relevant to modern ferrous metallurgists and their bulk process steelmaking techniques.

In his history of the development of iron-making in 17th and 18th century England and Wales, Hayman (2005), provides relevant commentary on the problem of technical vocabulary and the language of workmen.

2

The arcane glossary of terms to define aspects of furnace and forge was part of the industry mystique and helped to foster a conservative, inward-looking culture. The use of technical vocabulary in contemporary documentary sources is always at one remove from its source, since these documents were written by managers and ironmasters. Moreover, as an oral culture, the language was never definitively codified. For these reasons the use of terms can seem idiosyncratic and not always susceptible to definitive interpretation. Meanings also changed over time and so a term must be interpreted within its specific context. (Hayman 2005, 59)

Modern Science and Technology

The understanding of ancient steel- and toolmaking techniques is significantly enhanced by our ever-growing knowledge of the chemistry of modern ferrous metallurgy. No ancient smelters or edge toolmakers had any knowledge of the role of carbon in making steel or of the atomic structure of the world in which they lived. And yet, underlying our 21st century analytical understanding of chemistry and metallurgy are medieval and Renaissance technologies based on long established craft traditions. Our interpretation of early techniques and technologies derives from the projection of our contemporary knowledge of modern chemistry and the microstructure of ferrous metals on these earlier traditions. The growing diversity and complexity of alloy steels, recent advances in powdered metallurgy, an increasing understanding of superplasticity in metals, and the vast expansion of our understanding of the chemistry and kinetics of modern ferrous metallurgy assist us in reevaluating the older steel- and toolmaking strategies and techniques. The modern sciences of metallurgy and metallography thus serve as bridges connecting two continents; one is made up of lost or obsolete techniques and terminology, and the other consists of the vast amounts of information now available through modern information and communications technologies.

Shortcomings of Modern Analytical Chemistry

Modern science has yet to unravel and explain in rational scientific language many of the complexities of both ancient and modern ferrous metallurgical techniques and the phase transformation kinetics upon which they are based. Contemporary iron-carbon diagrams are sufficient only for simple schema. Modern science falls well short of being able to explain adequately the art of edge tool production, given the wide variety of micro-constituents as alloys (deliberately added) or micro-contaminants (accidentally present), wide variations in the time and temperature of thermal and mechanical treatment processes, and multiple strategies used in attempts to forge edge tools. In particular, the complexity of the microstructure of the many varieties of iron and steel have not yet been fully deciphered, despite the growing capacity of nanotechnology to penetrate this maze. For example, there is no modern text that adequately explains the techniques and scientific basis for the ancient production of Damascus steel swords. In this context, words, including glossary definitions, are inadequate to explain the marriage of art, tools, and history.

Given the challenges of interpreting the long history of steel- and toolmaking strategies and techniques, the author reviewed numerous antiquarian and contemporary sources to compile accurate and complete definitions of terms pertaining to the history of ferrous metallurgy. The revised second edition also includes many new definitions and descriptions of modern carbon, alloy, tool, and stainless steels, as well as a number of riffs on the challenges faced by born-again ironmongers as contemporary tool wielders in the context of an increasingly dysfunctional contemporary consumer society now entering the early stages of post-Fukushima Daiichi, post-Sandy age of biocatastrophe. Please contact the author with corrections, suggested revisions, comments, or additional definitions at curator@davistownmuseum.org.

On Ironmongers

The *Oxford English Dictionary* has the following definition of "monger":

> In formations dating from the middle of the sixteenth century onward monger nearly always implies one who carries on a contemptible or discreditable "trade" or "traffic" in what is denoted by the first element of the compound, as in fishmonger… ironmonger.

In European communities before the settlement of North America and in colonial America and the early republic, ironmongers were ironworkers following the trades of the bloomsmith, forge master, founder, shipsmith, edge toolmaker, farrier, or community blacksmith. Ironmongers usually worked in darkened environments, often at night, where the lack of sunlight allowed them to make the critical judgment of whether the iron implements and steel tools they were forging or repairing were ready for quenching and tempering or other heat treatments, as shown by the cherry red color of steels heated to the proper temperature for forging (1375° F). Before the science of metallurgy and the chemistry of steel as an iron carbon alloy were known, ironmongers were considered to be alchemists who somehow possessed the fiery secrets of forging the iron and steel tools essential to the survival of every maritime and horticultural community. In the age before the appearance of Lava soap, ironmongers were dirty and sooty, often too sooty, in fact, to attend Sunday church services. Their odd hours, habits, and alchemical finesse reinforced the perception of their peculiarity as metallurgical magicians, who were often so suspiciously regarded that, in African communities at least, they were required to live in homes located outside of the villages in which they worked.

When the first North American colonial integrated ironworks was established at Saugus, MA, in 1646, Joseph Jenks and other toolmakers who worked at this facility were emblematic of the commencement of a vigorous colonial iron industry that would play a major role in the success of the American Revolution. The finesse of colonial bloomsmiths, shipsmiths, and edge toolmakers was, in turn, based on two millennia of Celtic ferrous metallurgy. This glossary is an attempt to retrieve and recapitulate information about the forgotten steel- and toolmaking strategies and techniques of these earlier centuries and their legacy to the modern science of ferrous metallurgy.

Time Line

Table 1. The following time line provides a brief summary of the chronology of important events pertaining to the steel- and toolmaking strategies of New England's shipsmiths and edge toolmakers. Descriptions with an * are excerpted from Barraclough (1984a).

Date	Event
1900 BC	First production of high quality steel edge tools by the Chalybeans from the high quality iron sands of the south shore of the Black Sea.
1200 BC	Steel is probably being produced by the bloomery process.*
800 BC	Carburizing and quenching are being practiced in the Near East.*
800 BC	Celtic metallurgists begin making natural steel in central and eastern Europe.
650 BC	Widespread trading throughout Europe of iron currency bars, often containing a significant percentage of raw steel
400 BC	Tempered tools and evidence for the 'steeling' of iron from the Near East*
300 BC	The earliest documented use of crucibles for steel production was the smelting of Wootz steel in Muslim communities (Sherby 1995a).
200 BC	Celtic metallurgists begin supplying the Roman Republic with swords made from manganese-laced iron ores mined in Austria (Ancient Noricum).
55 BC	Julius Caesar invades Britain
50 BC	Ancient Noricum is the main center of Roman Empire ironworks. Important iron producing centers are also located in the Black Mountains of France and southern Spain.
43-410 AD	Romans control Britain.
125 AD	Steel is made in China by 'co-fusion'.*
700	High quality pattern-welded swords being produced in the upper Rhine River watershed forges by Merovingian swordsmiths from currency bars smelted in Austria and transported down the Iron Road to the Danube River.
1000	First documented forge used by the Vikings at L'Anse aux Meadows (Newfoundland)
1350	First appearance of blast furnaces in central and northern Europe
+/- 1465	First appearance of blast furnaces in the Forest of Dean (England)
1509	[Natural] steel made in the Weald [Sussex, England] by fining cast iron*
1601	First record of the cementation process, in Nuremberg*

Date	Event
1607	First shipsmith forge in the American colonies used at Fort St. George, Maine
1613/1617	Cementation process is patented in England.*
1617-1619	The great pandemic sweeps through the indigenous communities of the New England coast east of Narragansett Bay.
1625	First Maine shipsmith, James Phipps, working at Pemaquid
1629-1642	The great migration of Puritans from England brings hundreds of trained shipwrights, shipsmiths, and ironworkers to New England.
1646	First colonial blast furnaces and integrated ironworks are established at Quincy and Saugus, Massachusetts.
1652	James Leonard establishes the first of a series of southeastern Massachusetts colonial era bog iron forges on Two Mile River at Taunton, Massachusetts.
1675-1676	King Philip's War in southern Massachusetts and Rhode Island
1676	The great diaspora (scattering) of Maine residents living east of Wells follows the King Philip's War
1686	First documented use of the cementation process in England
1689-1697	The war of the League of Augsburg
1702-1714	The war of Spanish Succession
1703	Joseph Moxon ([1703] 1989) publishes *Mechanick Exercises or the Doctrine of Handy-Works*.
1709	Abraham Darby discovers how to use coke instead of coal to fuel a blast furnace.
1713	First appearance of clandestine steel cementation furnaces in the American colonies
1720	First of the Carver, Massachusetts blast furnaces established at Popes Point
+/- 1720	William Bertram invents manufacture of 'shear steel' on Tyneside.*
1722	René de Réaumur (1722) provides the first detailed European account of malleableizing cast iron.
1742	Benjamin Huntsman adapts the ancient process of crucible steel-production for his watch spring business in Sheffield, England.
1754-1763	The Seven Years War, the last of the French and Indian wars
1758	John Wilkinson begins the production of engine cylinders made with the use of his recently invented boring machine.
1759	The defeat of the French at Quebec by the English signals the end of the struggle for control of eastern North America.
1763	The Treaty of Paris, which opened up eastern Maine for settlement by English colonists

Date	Event
1763-1769	James Watt designs and patents an improved version of the Newcomb atmospheric engine, i.e. the steam engine.
1774	John Wilkinson begins the mass production of engine cylinders used in Watt's steam engine pressure vessels.
1775	Matthew Boulton and James Watt begin mass production of steam engines.
+/-1783	The approximate date when Josiah Underhill began making edge tools in Chester, NH. The Underhill clan continued making edge tools in NH and MA until 1890
1783	James Watt improves the efficiency of the steam engine with introduction of the double-acting engine.
1784	Henry Cort introduces his redesigned reverbatory puddling furnace, allowing the decarburization of cast iron to produce wrought and malleable iron without contact with sulfur containing mineral fuels.
1784	Henry Cort invents and patents grooved rolling mills for producing bar stock and iron rod from wrought and malleable iron.
1789-1807	Era of great prosperity for New England merchants due to the neutral trade
1793	Samuel Slater first began making textiles
1802-1807	Henry Maudslay invents and produces 45 different types of machines for mass production of ship's blocks for the British Navy.
1804	Samuel Lucas of Sheffield invents the process of rendering articles of cast iron malleable.
1813	Jesse Underhill is first recorded as making edge tools in Manchester, NH.
1815-1835	The factory system of using interchangeable parts for clock and gun production begins making its appearance in the United States.
1818	Thomas Blanchard designs a lathe for turning irregular gunstocks.
1820	Steam-powered saw mills come into use near Bath, Maine, shipyards.
1828	Adoption of the hot air blast improves blast furnaces
1831	Seth Boyden of Newark, NJ, first produced malleable cast iron commercially in the US.
1832	D. A. Barton begins making axes and edge tools in Rochester, NY.
1832-1853	Joseph Whitworth introduces innovations in precision measurement techniques and a standardized decimal screw thread measuring system.
1835	Malleableized cast iron is first produced in the United States.
1835	Steel is first made by the puddling process in Germany.*
1835	The first railroad is established between Boston and Worcester, Massachusetts.

Date	Event
1837	The Collins Axe Company in Collinsville, Connecticut, begins the production of drop-forged axes.
1837	In England, Joseph Nasmyth introduced the steam-powered rotary blowing engine.
1839	William Vickers of Sheffield invented the direct conversion method of making steel without using a converting furnace.
1842	Joseph Nasmyth patents his steam hammer, facilitating the industrial production of heavy equipment, such as railroad locomotives.
1849	Thomas Witherby begins the manufacture of chisels and drawknives in Millbury, MA.
1850	Joseph Dixon invented the graphite crucible used in a steel production.
1853	John, Charles, and Richard T. Buck form the Buck Brothers Company in Rochester, NY, after emigrating from England and working for D. A. Barton. They later move to Worcester, MA in 1856 and Millbury, MA in 1864.
1856	Gasoline is first distilled at Watertown, Massachusetts.
1856	Bessemer announces his invention of a new bulk process steel-production technique at Cheltenham, England.*
1857	The panic and depression of 1857 signals the end of the great era of wooden shipbuilding in coastal New England.
1863	First successful work on the Siemens open-hearth process*
1865	Significant production of cast steel now ongoing at Pittsburg, Pennsylvania furnaces
1868	R. F. Mushet invents 'Self-hard,'the first commercial alloy steel.*
1870-1885	Era of maximum production of Downeasters in Penobscot Bay (large four-masted bulk cargo carriers)
1874	Tilting band saw is introduced and revolutionizes shipbuilding at Essex, MA.
1879	Sidney Gilchrist Thomas invents basic steelmaking.*
1906	The first electric-arc furnace is installed in Sheffield.*
1913	Brearley invents stainless steel.*
1926	The first high-frequency induction furnace in Sheffield*

*(Barraclough 1984a, 13-4).

Handbook for Ironmongers: A Glossary of Ferrous Metallurgy Terms

The definitions in this text derive from a wide variety of contemporary and antiquarian sources pertaining to the art of the bloomsmith, forge master, blacksmith, edge toolmaker, shipsmith, founder, and other artisans of ferrous metallurgy. A bibliography of sources used in the compilation of this glossary follows the other appendices and is followed by a general bibliography on metallurgy. More comprehensive bibliographies pertaining to steel- and toolmaking strategies and techniques before 1870, the evolution of iron-smelting and toolmaking in colonial New England, and the florescence of American toolmakers are included in each volume of the *Hand Tools in History* series. All bibliographies are reprinted in Davistown Museum publication 48: *Davistown Museum: The Complete Bibliographies.* This handbook also includes a variety of definitions and notes pertaining to our contemporary (2013) social, political, economic, and environmental milieu and the challenges faced by tool wielding artisans and craftspersons (born-again ironmongers) as the tragedy of our round world commons unfolds.

Acicular: An adjective describing interlocking needle-like formations that characterize the crystal structure of ferrous metals under a variety of cooling conditions. Slowly cooled pearlite, formed during the production of blister steel, has acicular characteristics in the form of acicular ferrite. The rapid cooling of austenite results in the instantaneous formation of martensite, another allotropic form of iron characterized by an acicular crystal structure (Bain 1939). See allotropic forms of iron, blister steel, crystal structure, ferrite, and pearlite.

Acicular ferrite: A common component of pearlite, ferrite that has acicular crystal structures characterized by "tough and strong microstructures" (Sinha 2003, 9.24).

Acicular mode of transformation: Rapid quenching of austenite (eutectoid carbon steel) during a time period of two minutes or less to slightly above or near room temperature (200° F) results in transformation of austenitic microstructures by shearing mechanisms that produce martensitic structures. In contrast, slow cooling rates are characterized by carbon diffusion, which produces ferrite and cementite matrixes, i.e. pearlite. The crystallographic planes of martensite have an acicular structure as a result of their near instantaneous transformation; any remaining austenite also suffers deformation (Bain 1939). See acicular.

Acid process: Production of steel using furnaces lined by a siliceous refractory, such as sand or clay, suitable for low phosphorous ores only. Until 1877, the acid process was the primary means of bulk process steel production. Introduction of the basic process after 1877 allowed smelting of ores with a high phosphorous content. See basic process and ganister.

Acid steel: Steel melted under a slag that has an acid reaction and in a furnace with an acid bottom or lining (Shrager 1949).

Acier fondu: The French term for cast steel.

Acier forge: The French term for fined (decarburized) pig iron. See pig iron.

Adz (adze): A common woodworking tool used for smoothing wood both across and with the grain. The most ancient and important of all forms of woodworking tools, adzes differ from axes in that the adz head is aligned with the haft, like a hoe. Primitive forms of stone adzes were used as early as the beginning of the Stone Age. Cold hammered bronze adzes were the most important shipbuilding tool in the Bronze Age. The oldest known "steeled" adz is a malleable cast iron adz with a heat-treated steeled cutting edge made in China, c. 950 BC (Barraclough 1984a). Iron, natural steel, and steeled iron adzes were commonplace throughout the Iron Age and well documented in Roman archeological sites (Manning 1985). Goodman (1964) has excellent illustrations of both the socket and shaft hole adz forms. The most famous of all adzes is the unique New England "Yankee Pattern" ship carpenter's lipped adz, which was first made sometime in the 18th century and probably designed by some Gulf of Maine shipsmith whose identity has been lost. Also called a dubbing adz, this tool was used for shaping the frames and ribs of ships.

Age hardening: Changes in the physical properties of low carbon steels and other nonferrous metals at or slightly above room temperature over a period of time where unstable crystal structures gradually return to more stable forms.

Age of Bioengineering: See bioengineering, age of.

Air blast heat exchangers: In direct process Catalan forges, air blast heat exchanges facilitated the transfer of heat from the furnace directly to the air blast by the use of horizontal iron pipes that were exposed to furnace exhaust and connected to vertical iron pipes that lead into the tuyère. Various designs of air blast heat exchangers characterize all modern blast and cupola furnaces.

Air furnace: A furnace used to melt cast iron and to refine other metals. The fuel, which is usually coal or coke, and the flame that passes over the bath holding the melted metal, do not come in contact with the metal, and, thus, the carbon content of iron can be carefully controlled. The term air furnace is synonymous with reverbatory furnace. See reverbatory furnace.

AISI: See American Iron and Steel Institute.

Albedo: the reflection of sunlight from polar ice and snow cover; also the reflection of high frequency ultraviolet light from the sun by the stratospheric ozone layer. Decreasing surface albedo is an important cause of global warming. A decrease in the earth's stratospheric ozone layer as a result of chlorofluorocarbon emissions increases the risk to human health from ultraviolet light.

Allotropic/Allotropy: Property of having changing physical characteristics, e.g. crystal structure, while retaining elemental chemical identity, i.e. an iron carbon alloy. Allotropy is expressed as the multiple forms of a space lattice in an element, i.e. its microstructure. See allotropic forms of iron, allotropic transformation, crystal structure, phase transformation, and space lattice.

Allotropic forms of iron: Iron has three allotropic forms. **Alpha iron**, also called ferrite, the low temperature form of iron up to 1670° F (910° C), has a body-centered cubic structure (BCC) and is soft, ductile, and magnetic. **Gamma iron**, also called austenite, with a face-centered cubic form (FCC), is nonmagnetic in a temperature range from 1670° F to 2540° F (1393° C) and is more compact and dense than alpha iron due to its carbon content being in a solid solution. In its pure iron form, austenite is also soft and ductile (Sinha 2003, 1.1). **Delta iron**, with a body-centered lattice structure (BCC) between 2240° F and 2793° F (1538° C), exhibits magnetism above its threshold temperature of 2240°F. The space lattice structure of iron breaks down above 2793° F (1538° C) as iron enters the liquid state. Temperature determines the crystal structure of ferrite-iron carbide combinations. Variations in carbon and alloy content, quenching, tempering, and cooling rates further influence the allotropic and microstructural forms of iron, steel, and cast iron. See the iron carbon diagrams in *Appendix VI*.

Allotropic transformation: Spontaneous and reversible changes in the microstructure of steel made by changing the temperature of the metal.

Alloy: Metallic substance fused with either another metal or a non-metallic element, e.g. steel is an alloy made of iron and carbon. The metals most commonly alloyed with iron are nickel, silicon, manganese, tungsten, chromium, cobalt, vanadium, and molybdenum. See brass and bronze.

Alloy steels and alloy cast irons: Steel and iron containing other elements in addition to carbon. Strength, durability, toughness, hardness, corrosion resistance, fracture, and carbon distribution are among the many properties determined by metals alloyed with iron and steel. Some alloying elements serve to increase the formation of martensite at slower rates of cooling or at lower temperatures, therefore increasing hardenability. In the AISI-SAE classification system, alloy steels are defined as any steel exceeding element limits of carbon steels, e.g. manganese 1.65%, silicon 0.9%, and copper 0.6%. See carburize, decarburization, and martensite.

Alpha iron: Another term for ferrite (pure iron); an iron with the face-centered cubic (FCC) form at temperatures below 1414° F (768° C). The allotropic form of iron begins changing to BCC at 1414° F, and the transition to Gamma iron is completed at 1670° F (910° C). Alpha iron has a carbon content ≤ 0.03% and is the softest component in the microstructure of steel. See carbon content of steels, ferrite, and wrought iron.

American felling ax: A distinctive form of American ax developed in the 18th century in response to the need for a felling ax with a heavier poll than those on traditional English felling axes. Such felling axes were usually forged out of two slabs of iron, which were steeled with an inserted bit after being folded and welded (Kauffman 1972).

American Iron and Steel Institute (AISI): An association of North American steel producers. They, along with the Society for Automobile Engineers (SAE) created the modern classifications of standard steel, alloy steel, stainless steel, and tool steel. See *Appendix IV*, carbon steel classifications, and classification system for steel.

American Society for Testing and Materials (ASTM): Recently renamed ASTM International. Established in 1898 by the Pennsylvania railroad, the ASTM now promulgates the standards for materials, including steel and iron used in industrial processes. See *Appendix IV*.

American system of manufacturing: Developed after 1830, the American system is characterized by the waterpower-driven production of tools and firearms with machine-made, interchangeable parts. The lower Connecticut River Valley was America's first center of manufacturing utilizing interchangeable parts. Eli Terry (clockmaker), Elisha Root (axes), Eli Whitney (firearms manufacturing), and John Hall (firearms manufacturing) are some of the major figures in the evolution of the American factory system. See the Tool Manufacturing Chronology online at http://www.davistownmuseum.org/TDMtoolHistory.htm#toolMfgChron.

Amisenian: Synonymous with Chalybean, the Anatolian community that produced high quality nickel steel from Black sea sands using chloanthite as a flux during the Bronze Age (1900 BC). See Chalybean steel.

Anchorsmith: Blacksmith who specialized in making anchors. Before 1800, anchors were often made from forged wrought iron produced by a bloomsmith at a direct process bloomery. From the early Iron Age until the 18th century, many shipsmiths were also anchorsmiths as well as edge toolmakers. After 1800, puddled iron gradually replaced direct process bog iron for anchor production, and trade specialization resulted in the separation of these crafts into individual professions. See blacksmith, bloomsmith, and shipsmith.

Ancony: Short thick wrought or malleable iron bar stock produced in a finery from shingled iron before being sent to the chafery for shaping into special sizes suitable for North American colonial era production of iron tools and artifacts. Anconies were two to three feet long with knobbed or square ends (Gordon 1996). See chafery, finery, malleable iron, and wrought iron.

Ångström: A non-SI (International System of Units) unit of measurement denoted by the symbol Å which is equal to 1/10,000,000,000th of a meter ($1 \times 10{-}10$ m).

Anisotropic: An adjective used to describe materials having different properties determined by variations in the directional arrangement of their crystallographic patterns. Directional dependency is also expressed by the related root word anisotropy. See crystal structure, grain, isotropic, and microstructure.

Anneal: To heat steel to a temperature, which is usually, but not always, below its critical temperature, followed by slow cooling to toughen it by altering its microstructure, which allows the uniform distribution of particles of cementite within the crystal structure of the annealed steel. Since, in many cases, annealed steel has not been quenched and tempered, annealing can be considered an alternative method of softening steel or otherwise relieving strain hardening. The variants of time, temperature, carbon, and alloy content allow an immense variety of annealing techniques. "...to relieve internal strains and improve strength, elasticity and ductility to meet the stresses to which it is subjected in service. It is usually done by heating and holding at a certain temperature followed by slow cooling" (Salaman 1975, 246). Annealing changes the diffraction patterns of the crystal structure of the metal being annealed. During the annealing of white cast iron, which typically takes 60 hours or longer, all the chemically combined carbon is reduced to free carbon or graphitic carbon surrounded by pure iron. Iron oxide or millscale packed with the white cast iron facilitates malleable cast iron production by assisting in the freeing of the carbon from its chemical bond within the iron. Many variations of the annealing process produce a wide variety of forms of malleable cast iron. Partially decarburized, annealed cast iron can be reheated and suddenly cooled, giving a steely fracture. In many cases, the specific procedures and techniques used to produce malleable and grey cast iron artifacts, such as Griswold cast iron cookware, were highly guarded trade secrets. In plain carbon steels, annealing usually results in the production of ferrite-pearlite microstructures. See full annealing, crystal structure, fracture, malleable cast iron, martensite, temper, and white cast iron.

Annealed eutectoid steel: Pearlite, the transformation product of austenite containing 0.83% carbon content, after *slow* annealing. Variations in the temperature and rate of annealing result in variations in the microstructure of the pearlite being formed as annealed eutectoid steel. The rapid cooling of austenite produces martensite. See annealing, austenite, eutectoid, martensite, and pearlite.

Annealed hypereutectoid steel: A pearlite-cementite matrix, the transformation product of austenite containing *more than* 0.83% carbon content after slow cooling. Variations in the temperature and rate of annealing result in variations in the microstructure of the pearlite-cementite matrix being formed as annealed hypereutectoid steel.

Annealed hypoeutectoid steel: A pearlite-ferrite matrix, the transformation product of austenite containing *less than* 0.83% carbon content after slow cooling. Variations in the temperature and rate of annealing result in variations in the microstructure of the pearlite-ferrite matrix being formed as annealed hypoeutectoid steel. The rapid cooling of austenite, followed by the termination of cooling between 800° F and 400° F and holding of the temperature for a variable time period, usually hours (austempering), creates microstructures known as bainite, the intermediary between ferrite-pearlite (uninterrupted slow cooling) and martensite (fast cooling, as with quenching). A scientific understanding of the microstructures of ferrous metals including eutectoid steels was not clearly elucidated until Edgar C. Bain (1939) published his classic text *Functions of the*

Alloying Elements in Steel. See bainite.

Annealing pots: Cast iron pots utilized to enclose and protect steel being annealed from the formation of iron oxide scale; used especially for annealing wire.

Annealing, purpose of: One of the most important strategies for producing a wide variety of modern steels. There are now dozens of annealing techniques used for the basic heat treatment of many special purpose alloy steels. Shrager (1949) has these comments on annealing steel:

> Annealing may be performed by one of several methods, depending upon the results desired. The purpose of annealing may be: (1) to remove stresses that have occurred during casting or as a result of work done on steel; (2) to soften steel for greater ease in machining, or to meet stated specifications; (3) to increase ductility in order to make steel suitable for drawing operations; (4) to refine the grain structure and make the steel homogenous; (5) to produce a desired microstructure. (Shrager 1949, 154)

All of Shrager's comments also pertain to the annealing of white and/or grey cast iron to produce the many forms of malleable cast iron. See grey cast iron, malleable cast iron, microstructure, stress, and white cast iron.

Antrim patent: Patent granted for welding steel straps onto the front band of shovels; used at Ames Shovel Works to produce the best quality shovels.

Anvil: "A heavy iron block with a hardened or applied steel topped face suitable for use by a blacksmith in working hot metal" (Sellens 1990, 5). In his *Dictionary of American Hand Tools*, Sellens lists 36 different types of anvils, many of which had a wide variety of forms, such as dengel stocks (scythe anvil). Most anvils had a flat work surface, in contrast to mandrels and other tools with curved work surfaces. Anvils were made from both wrought iron and cast iron and generally had a steel plate welded to the top. In the late 19th century, some makers produced solid cast steel anvils.

Apocalypse: The revelation of biocatastrophe; the disclosure, understanding, and experience of the synergistic components of biocatastrophe by the communities impacted by the disruptions of cataclysmic climate change, the hemispheric transport of ecotoxins and their bioaccumulation in pathways to human consumption, pandemics, and the infrastructure collapse that results as a consequence of the growth and demise of a non-sustainable pyrotechnic industrial society.

Archaeometallurgy: Study of ancient and antiquarian strategies and techniques for smelting, casting, and/or forging metals (copper, brass, bronze, lead, silver, iron, and steel) into hand tools and other artifacts; the study of the microstructure of metals produced by these techniques.

Archimedean screw: Hollow, inclined screw, usually in the form of a spiral pipe with an inclined axis. When used as a water pump, the lower end of the screw gathers in water, which is then discharged at the upper end; one of the first forms of the simple machine known as the screw. It was named after its Egyptian inventor, Archimedes (260 BC).

Arms race: Competition between cultures/countries to produce superior weapons. The first arms race occurred when the use of the blast furnace became widespread in northern Europe sometime during the late 14th and early 15th centuries. Its principal function was the production of cast iron cannon, which, along with bronze cannon, were used to arm the growing fleets of England, France, Spain, and The Netherlands. The growth of English merchant shipping and the Royal Navy were accompanied by the expansion of cannon and ordnance casting, and the proliferation of wrought iron handguns, such as the matchlock and then the flintlock. The arms race that began with the construction of the first blast furnace has continued unabated into the 21st century. All advances in ferrous metallurgy had and have as their primary social function the objective of improving weapons production. Successful exploration, conquest, and settlement of the New World were closely linked to innovations and improvements in the production of firearms.

Arquebus: The most common form of firearms at the time of Native American – European contact. The arquebus was always made of wrought iron and was brought to North America by French traders in exchange for furs, forever changing the lifestyles of the indigenous communities impacted by its availability. See matchlock.

ASTM: See American Society for Testing and Materials.

Atomic planes: Planes within a crystal, along which the atoms of a particular lattice structure are arranged.

Austemper: To produce bainite by the interrupted quenching of austenite in a bath of molten salt held at a temperature of 450° F - 800° F until bainite formation is completed. See bainite, martensite, and quench.

Austenite: Iron carbon alloy (steel), in which carbon is dissolved in iron as a stable solid carbon solution at a temperature range of 723° C (1400° F). Unstable at room temperature, austenite has a face-centered cubic (FCC) lattice structure, creating more space for a higher proportion of carbon to be held in interstitial solution than in the body-centered cubic (BCC) lattice structure of ferrite. If cooled slowly, granules of pearlite will appear in the austenite matrix, creating steel of inferior quality. If cooled rapidly, hard and brittle martensite will be produced, which, if then tempered (slowly heated and cooled), will be slightly softened and made less brittle by the homogenous redistribution of tiny spheroids of cementite, creating steel with a uniform carbon content (cc). See cementite and microstructure.

Austenite formation: Use of thermal treatments (heating) to change the microstructure and, thus, the allotropic form of a wide variety of ferrite-pearlite-cementite microstructures to austenite (gamma iron). Austenite formation occurs in two stages in low carbon steels: 1) pearlite transformation to austenite occurs at temperatures greater than 800° C; 2) ferrite to austenite transformation occurs at 910° C. If the latter temperature is not reached, "the second stage of further nucleation does not occur, and the process is complicated by the growth of the remaining austenite particles within the ferrite matrix" (Sinha 2003, 10.6).

Austenite nucleation sites: Since austenite is formed with pearlitic microstructures, three favorable nucleation sites are "[1] the ferrite/cementite interface, [2] the line of intersection between platelets and surfaces of the pearlite colony, and [3] the points of intersection between platelets and the edge of the pearlite colony" (Sinha 2003, 10.2).

Austenitic steels: Alloy steels containing chromium, nickel, or manganese that retain austenite at atmospheric temperatures.

Austenize: To create steel by heating iron above 723° C (1400° F) (the minimum austenizing temperature) and holding that temperature to allow the microstructural formation of austenite. The rate of cooling that follows determines the microstructural qualities of steel (Wayman 2000). Rapid cooling produces martensite; slow cooling allows austenite to return to ferrite and pearlite. See austenite, ferrite, martensite, pearlite, quench, and temper.

Austenizing temperature: In modern alloy steel production, the temperature between 1400° F (723° C) and 1900° F (1040° C), depending on carbon content (Pollack 1977). A more recent commentator notes austenite formation from various initial microstructures as starting at 735° C and being contingent upon various holding times (Sinha 2003, 10.1-.2). See austenize.

Ax (Axe): A common woodworking tool, on which the cutting edge, and, thus, the cutting angle, is in the same plane as the sweep of the implement in contrast to the adz, on which the cutting edge is perpendicular to the plane of the sweep.

Ax-Adz: Earliest metal form (copper) of a woodworking tool, dating to c. 4000 BC, 2100 years before the age of Chalybean steel. The cold hammered copper ax-adz was only slightly less dull than a stone ax. The cutting edges of the ax and the adz were perpendicular to each other as in a modern variation of this tool, one form of the fireman's ax.

Ax bar: Bar stock that is 1" x 3" and often provided in 12' lengths. A standard commodity of the 19th century, ax bar stock ranged from malleable bar iron suitable for steeling to German or blister steel suitable for additional steeling or forging. See Collins & Co., ax, ax-making techniques, eyepin, and overcoat method.

Ax-making techniques: Before the era of the one piece cast steel ax, three strategies for steeling an ax were used in most ax-making factories. The oldest method of steeling an ax, used since the early Iron Age, was to insert a piece of carbon steel between a folded slab of iron and then weld them together. A more modern form of steeling is the overcoat method, in which a piece of high carbon steel is wrapped or folded over the iron or low carbon steel ax body prior to welding. The third strategy for ax manufacturing is called the scarf-joint method and was primarily used on the Canada pattern broad ax. "The iron body was feathered to an edge and then the steel bit was scarfed onto the blade" (Lee 1995, 170). The modern method of drop-forging an ax involves cutting steel billets, typically into three pieces, and then heating them to a temperature of 2350° F prior to drop-forging and quenching in brine (Kauffman 1972, Klenman 1998). See ax, cast steel, steeling, billets, drop-forging, eyepin, and iron cored.

Babbitt: Alloy used to make bearings. In its original form, babbitt is 89% tin, 3.5% copper, 7.5% antimony, but it is also made with varying quantities of lead.

Bainite: Mixture of ferrite and cementite with a microstructure that forms during the rapid cooling of austenite to a temperature of 800° F, after which it is held for a long period of time at a temperature above 400° F, but less than 800° F. This quenching method is known as austempering. Bainite is an intermediary between ferrite-pearlite (slow cooling) and martensite (fast cooling, as with quenching) and is more ductile than pearlite but less brittle than martensite. The transformation of austenite to bainite results in a wide variety of microstructure formations and is contingent upon alloy content and cooling strategies, the latter of which may vary in duration, temperature, and cooling mediums, such as water, air, oil, or salt baths. Bainite is shown as ledeburite on some iron carbon diagrams. See annealed hypoeutectoid steel, austemper, austenite, cementite, ferrite, martensite, and pearlite.

Balkan Copper Age: The earliest pyrotechnic society, c. 6000 BC, the Balkan Copper Age culture evolved from the Neolithic Vinča culture, which was located along the shores of the Danube and flourished prior to the Copper Age in the Aegean. Early Iron Age smelting techniques and furnace designs can be traced back to this culture. Its most famous archeological site is Catul Huyuk (Çatalhöyük). See Renfrew (1973).

Ball vise: The principal vise used to hold die blanks for shaping drop-forged dies; also used by silversmiths and pewter spinners for securing die-sinking patterns for flatware, Britannia Ware, and pewter.

Bar mill: Mill in which billets of iron or steel are reduced in thickness and width by grooved rollers.

Basic oxygen furnace: Modern furnace used to make steel by melting scrap steel using oxygen produced in an oxygen plant as the fuel.

Basic oxygen process (BOP): One of the two most widely used steelmaking processes in the 20th century. The basic oxygen process is an offshoot of the Bessemer process, utilizing pure oxygen for the rapid decarburization of cast iron, with significantly less pollution produced as an unwanted byproduct. The basic oxygen steelmaking process was introduced in the US in 1955 and quickly supplanted open hearth furnace steel production, reaching almost 65% of total US steel production by 1980. Furnace design and the charges utilized, pig iron high in manganese content and low in phosphorous, are similar to the now obsolete Bessemer furnace. Oxygen steelmaking processes are subdivided into the top-blown oxygen process and the bottom blown process, with the recent evolution of numerous special subtypes, including some suitable for processing the high phosphorous ores that dominate European rock ore production. No tool steels are produced by basic oxygen processes.

Basic steelmaking process: Process of making steel in steel furnaces with linings of limestone and other basic elements. This process revolutionized bulk steel production processes by allowing the smelting of ores high in phosphorous, which had not been possible in furnaces with acid linings such as sand and clay. Invented by Sidney Gilchrist Thomas in the 1870s and patented in 1879, this process was particularly useful for the open-hearth and Bessemer process and furnaces in continental Europe, especially Germany, located near high phosphorous ore deposits.

BDTT: Brittle-ductile transition temperature.

Beak iron: Tools with various designs of horns, points, or beaks, which could be inserted into swage blocks and anvil hardy holes for the purpose of shaping iron and steel, especially sheet iron. Many European and American specimens were collected in Europe and brought to America by Kenneth Lynch (2007) and are illustrated in the publication *The Armourer and his Tools :The Kenneth Lynch Tool Collection*. Historically important variations of beak iron, all derived from European prototypes, include the blow horn and ball-top, flat-top, and other stakes, which were also first produced in the U.S. by the North Brothers in Connecticut.

Bears: Ferruginous furnace bottoms produced by the smelting of copper during the Copper and Bronze Ages, when iron oxide (Fe_2O_3) was used as a flux; also called salamanders. During the Iron Age, many of these bears, rich in iron oxide, were re-smelted to produce iron tools and weapons.

Beating out: Shaping of the curved futtocks and other components of the ship's frame with the use of the broad ax either in woods at the timber harvesting site or at the shipyard. The Yankee pattern broad ax is the generic ax used by New England shipwrights for this purpose. See shipsmith and broad ax.

Bed charge: The pig iron placed on the coke bed prior to firing a cupola furnace.

Belgian iron: Soft wrought iron preferred by gun makers in Europe and Britain during the Renaissance (1450-1650); more easily forged and welded than other irons and steels used in gun barrel forging (Greener 1910).

Bell metal bronze: 75-80% carbon, 20-25% tin. Tin serves as a hardener.

Bergman, Tobern: Swedish chemist who was among the first researchers to identify carbon as the critical component of steel in 1791.

Bessemer, Henry: (1813-1898). Inventor of the process of manufacturing low carbon steel by oxidizing liquid pig iron with an air blast, thereby burning out silicon and other constituents. Bessemer issued his famous paper *The manufacture of malleable iron without fuel* in England in 1856. Due to the problem of phosphorus in the ores used to make the pig iron, Bessemer was initially unsuccessful in producing a useful malleable iron or low carbon steel product due to over oxidation. Robert F. Mushet soon suggested the addition of spiegeleisen, an iron-carbon-manganese alloy, and after establishing a steelworks in Sheffield in 1858, Bessemer gradually perfected the art of the mass production of steel. See Mushet, Robert F.

Bessemer process: Conversion of pig iron to steel in one step; first described by Henry Bessemer in 1856. A hot air blast produces Bessemer steel in a closed converter by rapid oxidation of the impurities in the pig iron, including silicon and most carbon.

> When Bessemer and open-hearth steels made their appearance in the market, an attempt was made to use them instead of wrought iron as the base for high-grade crucible steels. Though seemingly pure enough, apparently purer even than wrought iron, these metals were not able to compete with wrought iron for this purpose. For some reason, not yet satisfactorily explained, these new materials, which are made in 15, 35 and 50-ton batches, when used as a base, do not give as high quality tool steel as puddled wrought iron, which is slowly and laboriously made in 500-pound lots. (Spring 1917, 121-122)

> The Bessemer process, which gave a brilliant fireworks display, was a rapid one, taking less than half an hour. For this reason, it was not easy to control the process, so the steel varied more than was thought proper. (Abell 1948, 148)

Until the addition of spiegeleisen to strengthen it, Bessemer steel was unsuitable for many of its later uses. See crucible steel, manganese, open-hearth process, silicon, spiegeleisen, and wrought iron.

Bickern: "A smaller, TANGED version of the ANVIL that was particularly useful for small, rounded objects that were too small to be worked on the large anvil horn" (Light 2007, 97).

Billets: Wrought or malleable iron bars forged by a blacksmith from an iron bloom. In the early Iron Age, billets were made from smelters' currency bars after the removal of slag and consolidation of iron. After the development of bulk process steel, billets are defined as semi-finished squares of iron and steel made from red-hot ingots prior to further mechanical (rolling and forging) and thermal treatment. Roman billets were typically 5 – 10 kg (Tylecote 1987).

Bimetallic tools: Axes with iron or steel cutting edges attached to bronze heads; axes with iron cores covered with bronze (1300 – 1000 BC) (Wertime 1982).

Bimetallism: In early polymetalic societies, the production of tools and weapons using more than one metal, e.g. pattern-welded bronze and iron knives (rare) or the use of bronze rivets in iron shafts for sword production.

Biocatastrophe: The simultaneous degradation of the earth's principal biomes and the consequential decline of the capacity of biome ecosystems to produce sufficient supplies of fresh water, affordable food, (including fresh and salt water fish), renewable biomass fuels, and other biome-derived natural resources to furnish the needs of billions of people dependent on the complex infrastructure grids of industrial society. Biocatastrophe is the future consequence of the synergistic interaction of overpopulation, cataclysmic climate and sea level change, oceanic pollution and fisheries depletion, deforestation, soil erosion, depletion of non-renewable energy sources, reduced biodiversity, fresh water depletion and contamination, regional and/or world wide pandemics, and the accumulation of biologically significant chemical fallout and other ecotoxins in pathways to human consumption. Political paralysis, social unrest, and entrenched income disparity will characterize a collapsing global consumer society where public needs far outstrip public resources. Iron-smelting, steel forging, and other pyrotechnic activities of industrial society and its transportation systems will continue to play a principal role in ecosystems degradation as important source points of biologically significant chemical effluents, increased CO_2 emissions, and non-renewable biomass fuel depletion. See mercury and infrastructure collapse.

Biocatastrophe in a nutshell: The synergistic impact of a lot of bad things happening all at once, i.e. the tragedy of the commons in a biosphere with finite natural resources.

Bioengineering, Age of: An optimist's name for the Age of Biocatastrophe.

Biologically significant chemical fallout (BSCF): Liquid, gaseous, and particulate emissions produced by the pyrotechnical activities of industrial society that have a deleterious impact on living organisms, including humans, and which play a hidden, but major, role in the ongoing decline of the biodiversity of most ecosystems. The proliferation of gene altering hormone disrupting chemicals (HDC) in domestic environments and consumer products (e.g. BPA) appears to be correlated with increases in autism spectrum disorders, birth defects, asthma, and possibly obesity. Methylmercury, dioxin, chlorinated hydrocarbons, perchlorates (rocket fuel), brominated fire retartants, and other persistent organic pollutants (POP) are among the ubiquitous byproducts of biomass fuels and their processing into thousands of chemicals used by industrial society or in its consumer products. Long-lived anthropogenic radioisotopes are the primary form of biologically significant chemical fallout produced by nuclear-powered heat engines (steam turbines). Contamination of ecosystems with biologically significant chemicals can take forms as diverse as stratospheric fallout (radioisotopes), tropospheric fallout (all forms of chemical fallout derived from evaporated or biogeochemically mobilized toxins), and liquid and/or particulate toxins emitted from terrestrial industrial and domestic environments and consumer products. Most forms of BSCF bioaccumulate in pathways to human consumption. See biocatastrophe, infrastructure collapse, mercury, and methylmercury.

Biomass fuels: Wood, peat, coal, coke, gas, and oil utilized as fuels for the heat engines of industrial society. Biomass fuels, most of which are nonrenewable, are a metaphorical bank account that, once depleted, cannot be replenished. See biocatastrophe, furnace types, and smelting.

Biosphere as bank account: Hundreds of millions of years of carbon cycling and photosynthesis created most of the biomass fuels used by industrial society to smelt and forge the earth's abundant iron resources into the infrastructure grids (e.g. cities and transportation systems, etc.) of contemporary global consumer society. Implicit in the depletion of the earth's biomass as bank account is the contraction and then collapse of growth-based, consumer-product-centered, industrial society.

Biosphere as industrial effluent cesspool: Worldwide iron-deposit-derived steel production is now rapidly passing 100 million tons per month. Exploitation of natural resources, such as iron, coal, oil, and gas, occurs in conjunction with the use of the biosphere as an industrial effluents cesspool. The rapid increase in industrial production is accompanied by a concurrent reduction in biodiversity and natural resource availability. Combustion gasses, including CO_2 emissions and biologically significant chemical fallout from industrial infrastructure activities, including consumer product production, have as their probable outcome the synergistic interaction of the multiple components of biocatastrophe, resulting in infrastructure collapse as the legacy of pyrotechnic industrial society.

Biscayne ax: Ax produced in the Biscayne area in Spain (Catalonia), the principal location of the manufacture of trade axes in the 15th and 16th centuries. These axes were brought to the Carolinas and Florida by the Spanish and also transported over the Pyrenees to be sold to the French. The French also brought them to North America, especially to the St. Lawrence River region and traded them to indigenous communities for furs, hence the name "trade ax." See Russell (1967) and Kauffman (1972).

Bit: Steel cutting edge of an ax. See ax-making techniques and over-coating.

Black-heart: Form of malleable iron annealed only for a short time, with the result that its fracture has a white rim with a black center. Most American-made malleable iron takes this form (Spring 1917). See malleable cast iron and white-heart.

Blacksmith: Artisan who forms hot iron into tools and hardware. In some cases, especially in the early Iron Age and as recently as the 19th century, blacksmiths smelted their own blooms of iron and forged tools and weapons out of their loup of hot iron by the various tricks of their trade, which included hammering, quenching, tempering, annealing, piling, reforging, pattern-welding, casting, burnishing, etc. From the early Iron Age to the modern era, most working blacksmiths obtained the iron they were forging from the bloomsmith, then, later, from the blast furnace via the finery, and, today, from bulk process steel furnaces.

Blacksmith trades: Ideally, the blacksmith, in a New England shipbuilding community, for example, was a multi-tasking ironmonger forging or repairing horticultural tools, oxen shoes, knives, ironware for shipbuilding, edge tools, and guns. In reality, most blacksmiths forging their hot iron had specialized trades: edge toolmaker, shipsmith, gunsmith, farrier, shovel maker, and general purpose village blacksmith.

Bladesmith: A contemporary term for knife-makers, among the most ancient of all ironmongering trades.

Blast: Machine that pumps air to the tuyères of the blast furnace. See blowing tubs and trompe.

Blast furnace: High shaft furnace developed in Europe after 1350 to smelt iron ore in the reducing atmosphere of the partial oxidation of charcoal (and after 1750, of coke). The liquid iron produced was cast into cannons, other utensils, or "pigs" for further refinement (decarburization) into wrought and natural iron and steel. In modern smelting processes, a flux of limestone is used to combine with and remove silicate and other contaminants as slag. The blast furnace was a significant improvement over the smaller direct process bloomery furnace because it more efficiently reduced the ore to its metallic state with a higher carbon content and lower melting temperature than bloomery iron. Larger quantities of iron were produced with significantly less iron oxide lost as slag, therefore allowing smelting of lower quality ores. A series of improvements (including the use of the steam engine to enhance [cold] air blast [1775], furnace design changes [1784]], and the adoption of the hot air blast [1828]) increased operating temperatures, furnace capacity and efficiency, and the control of carbon content. New England's first blast furnace began operation at Saugus, Massachusetts, in 1646. See bloomery, arms race, cupola furnace, foundry, molding, Saugus Ironworks, slag, and carbon content of iron.

Blister steel: Steel produced in a cementation furnace, whose design protected the iron bar stock being carburized from oxidation caused by the burning fuel. Bars of wrought or malleable iron were stacked in the sandstone furnace with layers of charcoal dust or other carboniferous materials and fired for a period of five to twelve days. The resulting steel bar stock had a heterogeneous carbon content (cc) with less carbon in the inner layers, a higher carbon content in the outer layers, and blisters on the steel surface that were due to gas formation from the chemical reaction between the carbon and impurities in the slag. The slow cooling inherent in blister steel production resulted in the formation of ferrite-pearlite microstructures, including acicular ferrite. These microstructures required reheating and reforging followed by tempering for the formation of the more desirable allotropic forms of steel such as martensite-bainite, martensite-ferrite-bainite, and martensite-ferrite-pearlite-bainite matrixes, which are temperature- and transformation-time-dependent microstructures. (See the transformation time diagram in *Appendix VI*). Common 18th century and early 19th century names for these forms of reprocessed blister steel included designations such as spring, shear, and double shear.

Benjamin Huntsman's innovative introduction of easily hot-rolled cast steel (1742) revolutionized edge tool manufacturing by producing steel with uniform carbon content, which gave it characteristics of superplasticity at high temperatures and enhanced plasticity at room temperature, hence its durability and ease of sharpening. Blister steel bar stock was broken into pieces and re-melted in crucibles at high temperatures. The addition of carboniferous material (e. g. powdered charcoal) resulted in a chemical reaction that produced small quantities (5 - 50 kg) of cast steel with a highly homogenous distribution of carbon.

Blister steel was often used as a "weld" steel in the manufacturing of edge tools and was the principal form of steel produced in England from 1686 to 1850. Bars of blister steel were frequently piled (bundled) and reforged into higher quality, special purpose steel such as shear steel, double shear steel, and spring steel for saws. The edge toolmaker's act of forge welding a steel cutting edge to an iron pole or shaft also served the function of altering the microstructure of the steel, producing a more ductile or a harder steel cutting edge depending on the choice of mechanical and thermal treatments.

After 1750, despite the appearance of crucible steel, which was only produced in small quantities, blister steel, along with German steel, continued to dominate the world steel market until the development of bulk process steel (1870). See Bessemer steel, carbon content, cast steel, cementation furnace, German steel, open-hearth furnace, sap, shear steel, Sheffield classification of blister steel, spring steel, superplasticity, and wrought iron.

Bloom: **1**) (1870 definition) Partially melted, pasty mass of iron and slag resulting from the smelting of iron ore. Some bloomery-produced iron was nearly pure ferrite with a very low carbon content +/- 0.03% cc, but not more than 0.08% cc; slight changes in the fuel-ore ratio, tuyère location, and smelting temperature could produce iron with a variable carbon content ranging from that of malleable iron, with a range of carbon content from 0.08 - 0.5%, to raw steel, with a carbon content between 0.5 - 1.5%. Unwanted inclusions of pins and globules of steel would reduce the malleability and ductility of wrought and malleable iron. They often characterized the heterogeneous bloom produced in the direct process smelting furnace. These steel inclusions in the iron bloom may have also assisted blacksmiths in forging natural steel tools in early direct process open-hearth furnaces by encouraging experiments with furnace design and smelting processes; otherwise, steel nodules were considered to lower the quality and workability of wrought and malleable iron. **2**) (Modern definition) Hot-rolled, semi-finished billets, bars, and slabs of iron and steel produced in rolling mills. See carbon content of iron, direct process, ferrite, loup, tuyère, and wrought iron.

Bloom size: Early Iron Age blooms (Halstadt, La Téne, Noricum) were two to ten kilograms in size; medieval and late medieval bloomeries with water-powered trip hammers made larger blooms, 25 – 50 kg (Tylecote 1987).

Bloom smelting: Production of a loup of wrought or malleable iron from the reduction of iron ore (iron oxide) by the bloomsmith, whose success depended on accurate control of furnace temperature and the intuitive control of the gas ratio, i.e. CO/CO_2 (as in $C + CO_2 = 2CO$). Excess production of carbon monoxide during the smelting process would tend to recarburize the bloom of wrought or malleable iron, either accidentally or deliberately, producing higher carbon malleable iron, natural steel or unwanted cast iron.

Bloomery: Charcoal-fired shaft or bowl furnace used for the direct reduction of iron ore to produce wrought and malleable iron and natural steel. Bloomeries were the primary facility used to produce iron or, in some cases, raw steel before the appearance of the blast furnace (1350). They were in wide use in the United States until the Civil War and continued in use until the late 19th century in the United States and in some sections of

Europe, such as Catalonia, despite the proliferation of blast furnaces after 1600.

Bloomery iron: "'Bloomery iron' was essentially similar to 'wrought iron' in being virtually carbon free, but with entrained slag" (Barraclough 1984a, 11). Implicit in Barraclough's definition is a difference between the relatively slag-free wrought iron produced in the puddling furnace and the high slag content iron produced in the bloomery furnace. In New England, implements made from direct process bloomery-smelted bog iron ores have higher slag content than the wrought or malleable iron produced by decarburizing cast iron in the puddling furnace and are characterized by highly visible linear filaments of iron silicate. See bloomsmithing, malleable iron, silicon, wrought iron, and natural steel.

Bloomery steel: Steel produced either deliberately or accidentally by bloomsmiths operating a direct process furnace. Bloomery steel could appear as nodules of steel in the form of irritating contaminants in wrought or malleable iron bar stock or be deliberately produced by early Iron Age bloomsmiths, whose specific function was to make natural steel for swordsmiths and/or edge toolmakers. Bloomery steel was often beaten into thin sheets and piled in alternating layers with sheet wrought iron by swordsmiths, who made the ubiquitous pattern-welded swords from the early Iron Age to the late Renaissance. See natural steel.

Blooming mill: Modern term for rolling mills consisting of two high continuously reversing electrically driven rollers, which reduce steel slabs into appropriately-sized sheets, strips, and plates. Three high rolling mills require no reversing. In 1784, Henry Cort designed and built the modern form of the grooved rolling mill to shape puddled iron. See rolling mill.

Bloomsmith: Traditional name of the operator of a direct process bloomery from the early Iron Age until the late 19th century. The bloomsmith smelted iron ore, making the loup of wrought or malleable iron that would then be shaped by hand hammering, or, after 1350, by use of a water-powered helve hammer to expel the slag and other contaminants. In the early Iron Age, the bloomsmith would produce currency bars that were widely traded. After 1350, the use of the helve hammer allowed the bloomsmith to shape wrought or malleable iron bar stock. Both forms were then supplied to blacksmiths, shipsmiths, and swordmakers as the raw material used in their often bellows-driven forges. See currency bar.

Bloomsmithing: Direct process production of wrought and malleable iron by the hammering of the bloom at red heat to remove slag. It was an inefficient process due to the loss of +/- 50% of the iron content in the slag. In the early Iron Age, the first iron furnaces were run by bloomsmiths, who also functioned as blacksmiths, forge welding the first iron and steel tools. Historically, the bloomery process was the only practical way to obtain wrought and malleable iron until the widespread production of cast iron required strategies for decarburizing the cast iron. In the North American colonial period, bloomsmiths used bog iron, widely available along the coastal plains of the colonies, for the production of wrought and malleable iron. Successful bloomsmithing required a temperature of at least 1200° C to liquefy the slag that acts as a protective covering for the loup of iron being smelted. See carbon content of iron, bloomery, direct process, bloom, bloomsmith, bloom smelting, and natural steel.

Blow hole: Defect caused by trapped gas within molten metal; an ongoing problem in crucible steel production until the invention of the dozzle and also a continuing challenge of modern steelmaking techniques. See crucible steel and dozzle.

Blowing devices: Instruments used to provide the air flow or blast in bloomery and blast furnaces. The earliest blowing devices were leather bag bellows and fan bellows used in Asia. These later evolved into piston bellows. The trompe is a form of hydraulic bellows commonly used in Europe. The steam-powered blower was the universal blowing device used after 1775 to provide the air blast for modern blast furnaces before the advent of gas and electric motors. See blast furnace, blowing tubs, and trompe.

Blowing engine: Steam-powered rotary blowing engine, which was first introduced by Joseph Nasmyth in England in 1837 and greatly increased the operating temperature and efficiency of the blast furnace. One of a series of important innovations in blast furnace operation, it was quickly adopted by American ironmongers. See blast furnace and tuyère.

Blowing tubs: Wooden cylinders containing leather edged wooden pistons driven by water-powered piston rods (Gordon, 1996) of various and gradually improved designs based on piston-valve-cylinder combinations. Blowing tubs replaced bellows in the late 18th century, creating a stronger air blast in both bloomeries and blast furnaces. Smaller blacksmith and shipsmith forges continued to obtain air blast from the use of hand-powered accordion bellows until the end of the 19th century. See blast, blast furnace, and bloomery.

Bluing: The color of surface oxide on steel resulting from low temperature heat treatment (+/- 300° C).

Body-centered cubic structure (BCC): Low temperature (< 910° C / 1670° F) crystalline form of (alpha) iron in which atomic centers are located at the corners and center of sets of cubic cells; one of the three most common forms of crystal structures in most metals. After heating beyond the eutectoid temperature (723° C), the lattice structure of iron deforms into a face-centered cubic structure (FCC), as in gamma iron; this transformation is completed at a temperature of 910° C. At a temperature of (2552° F / 1400° C), gamma iron reverts to a BCC structure. The third form of cubic space lattice is the close packed hexagonal (CPH), characteristic of such non-ferrous metals as cobalt, magnesium, and titanium. Metals in this last group lack plasticity (Shrager 1949, 5-6). See lattice structure.

Bog iron: Iron precipitated in upwelling waters by contact with oxygen and organic material. In bogs and swamps, the constant process of bog iron formation can be seen by the characteristic brown scum on the top of the water formed by bacteria precipitating the iron from the water, which then falls to the bottom of the pond or swamp forming bog iron deposits. Limonite and goethite are the most commonly encountered hydrated iron ores in the North American coastal plain. Because both are combined with water, they are called hydrated iron ores. In the North American colonial era, bog iron was the most important local source of iron until the exploitation of mined terrestrial ores occurred around Salisbury, CT, in the early eighteenth century, as well as in Pennsylvania, New York, Maryland, and elsewhere. Pennsylvania soon became colonial America's most important source of terrestrial ore. High in phosphorous, bog iron, unless extensively refined, was smelted and forged into tools with highly visible slag inclusions. See bloomsmithing, bloomery, wrought iron, limonite, goethite, and hydrated.

Bombard: Earliest form of firearms utilizing gunpowder. The production of bombards in Europe after 1350 greatly increased the demand for wrought iron, which, in turn, facilitated the development of the blast furnace because it could produce large quantities of pig iron to be easily refined into wrought iron. See arms race, blast furnace, pig iron, wrought iron, and gunsmithing.

Born-again ironmonger: Individual who executes the convivial use of hand tools while at the same time resurrecting the knowledge of the ancient traditions of the ironmonger and the history of technology and its cascading Industrial Revolutions; the basis for the successful functioning of the skilled artisans and craftspersons essential for the viability of sustainable economies in the unfolding Age of Biocatastrophe.

Bosh: Tapered combustion zone of a blast furnace, located just above the hearth.

Botting: Plugging of the tap hole in a cupola furnace after removal of molten iron prior to the accumulation of the next batch of melted cast iron.

Bottom swage: Lower die placed into the hardy hole of an anvil for the purpose of shaping iron bar stock.

Bowl furnace: Most ancient form of an open-hearth furnace. The first bowl furnace was probably a campfire in which copper ore was accidentally melted. See furnace types.

Brass: Copper-zinc alloy with +/- 2.0% zinc content and lead as a micro-constituent.

Brazing: The process of joining of two pieces of metal by fusing a layer of brazing solder alloy (copper-zinc, copper-zinc-silver, or nickel-silver) between adjoining surfaces with two different melting points.

Breakdown furnace: Another name for a non-slag-tapping furnace where, at the end of the smelt, the furnace walls must be broken down in order to remove the bloom of iron or raw steel. Among the first forms of furnace design in the early Iron Age, open-hearth breakdown furnaces are still used in the early 21st century for production of raw steel for samurai swordsmiths.

Brescian steel: Ancient North Italian process for producing steel from carburized wrought iron soaked in molten cast iron; one of several steel producing technologies of the Italian city states during the Renaissance; later used in Europe and possibly in the

United States before the introduction of crucible steel. The cast iron was broken into pieces and mixed with saline marble as a flux and then melted. Billets of wrought iron were then placed in the molten cast iron for +/- 5 hours, removed, forged, and quenched, producing a high slag content steel with a variable carbon content. A similar procedure was also used to make steel in China, at least as early as 700 BC and probably earlier (Needham 1958, Barraclough 1984a). See Chinese steel.

Brine quenching: Expedition of uniform cooling by the chemical action of salt crystals; most preferred quenching medium of many early edge toolmakers.

Brinell hardness: Older method of determining hardness by measuring the diameter of an impression made by a ball of a given diameter under a specific load.

Britannia metal: Combination of tin, copper, and antimony to imitate pewter; invented in Sheffield, England, by James Vickers in the late 18th century (Tweedale 1986). Also called Britannia Ware.

British thermal unit (BTU): Amount of heat necessary to raise the temperature of one pound of pure water by one degree Fahrenheit starting at its maximum density temperature -39.1° F.

Brittle cleavage: Nearly instantaneous slip of the crystalline lattice structure of a metal.

Brittle-ductile transition temperature (BDTT): "...temperature at which iron upon cooling; suddenly becomes brittle. The transition temperature is sensitive to the purity of the metal" (Gordon 1996); also, the temperature at which iron and steel undergo a loss of toughness, usually below room temperature. Phosphorous raises BDTT above room temperature in the presence of carbon (Kauffman 1972). The most notable example of brittle-ductile transition temperature is the rapid cooling of austenite to form martensite. See cold short.

Brittleness: Tendency of a metal to suddenly fracture under low stress without deforming.

Broad ax: Form of ax with a wide blade (in excess of five inches); made in numerous styles ranging from the bearded broad ax of German origin to the wide Pennsylvania style broad ax. The "New England" style broad ax and the English pattern mast ax are the two most common forms of broad ax found in New England shipyards. The former was particularly important for use in the woods to "beat out" the white oak futtocks and keel components for New England shipwrights.

Bronze: Copper-tin alloy; +/- 10% tin.

Bronze edge tools: Most common form of woodworking tools before the Iron Age. Hammering flattened the cutting edge of a bronze tool, while, at the same time, hardening it. Bronze celts were often secured in knee bent hafts, later followed by enclosed hafts and socket cores. Bronze axes were frequently cast in molds with the aid of eyepins (Tylecote 1976). See cold hardened, celts, eyepin, and shaft hole vs. socket hole.

Bulat: Russian name for Wootz steel sword blades.

Bulk carburization: Production of steel in the order of magnitude of tons; first made possible by the design and use of the cementation furnace >1600. See basic steelmaking process, converting furnace, blister steel, Bessemer process, and Siemens-Martins process.

Butt riveting: Riveting of two plates of low carbon steel or iron together by the use of a covering plate that overlaps the two plates; one of two methods for building iron and steel ships. See lap riveting.

CAD: See computer-aided design and CNC technology.

Calcining: Roasting of ore to remove impurities and moisture. The resultant product was called ironstone, which was then smelted and reduced to loups (blooms) in a bloomery furnace. See bloomery.

Calescent: Characterized by increasing temperature, i.e. glowing with heat.

Cannon metal: Alloy of copper and tin, also called gun metal. See alloy and gunmetal bronze.

Carbide: Compound of carbon with one or more elements, i.e. iron carbide, Fe_3C; by definition, the added element(s) must be characterized by a lesser electro-negativity. See carbide forming metals.

Carbide banding: Patterns of aligned bands of cementite produced by the thermal and mechanical treatment of steel.

Carbide forming metals: Metals, which, when present at $20 - 100$ parts per million as micro-segregated components of the interdendritic regions of iron carbide microstructures, facilitate the formation of cementite bands in the form of nanowires and nanotubes during, for example, the forging of Damascus and samurai sword blades. Vanadium, chromium, titanium, molybdenum, niobium, and manganese are carbide forming metals. These metals, as alloys in steel, have a wide range of applications, from edge tool and knife-forging to manufacturing hardened steel cutting tools and shock-resistant turbine blades. See alloys, cementite, Damascus sword, nanotubes, nanowires, and superplasticity.

Carbide lamellae dimensions: "In the clusters of lamellae the films of hard carbide may be as thin as half of one millionth of an inch or perhaps as thick as 0.00008. In the case of spheroidal particle size the range is from what might be called molecular size to spheroids perhaps nearly 0.001 inch in diameter" (Bain 1939, 22-3). All size dimensions are controlled by the cooling and reheating of the steel, i.e. heat treatment cooling rates.

Carbon: Most important element in the creation of steel or cast iron from pure iron. The amount of carbon added to iron and its pattern of distribution as an alloy determines the microstructure and mechanical properties of steel or cast iron. With respect to the ferrous metallurgy of woodworker's edge tools, carbon is the key element differentiating iron from steel. Various forms of steel contain an intermediate level of carbon. Vandermonde, Berthollet, and Monge first identified carbon as a component of steel in 1786 (Partington

1961). Also see Bergman (1781).

Carbon content of ferrous metals: Sources vary widely in defining the *minimum* carbon content of steel, which ranges from 0.1 to 0.5% carbon. Please note the caveats that follow the definitions.*

Wrought iron: 0.01 – 0.08% carbon content (cc); soft, malleable, ductile, corrosion-resistant, and containing significant amounts of siliceous slag in bloomery produced wrought iron, with less slag in blast-furnace-derived, puddled wrought iron. Wrought iron is often noted as having ≤ 0.03% carbon content.

Malleable iron 1): 0.08 – 0.2% carbon content (cc); malleable and ductile, but harder and more durable than wrought iron; also containing significant amounts of siliceous slag in bloomery produced malleable iron, with less slag in blast furnace derived, puddled malleable iron.

Malleable iron 2): > 0.2 – 0.5% carbon content (cc). Prior to the advent of bulk-processed low carbon steel (1870), iron containing the same amount of carbon as today's "low carbon steel" (see below) was called "malleable iron." Its siliceous slag content gave it toughness and ductility, qualities not present in modern low carbon steel, hence its name. Before 1870, a wide variety of common hand and garden tools and hardware were made from malleable iron with a significantly higher carbon content than wrought iron.

Natural steel: 0.2% carbon content or greater. Natural steel containing less than 0.5% cc is synonymous with the term malleable iron. Natural steel is produced only by direct process bloomery smelting and was the only form of steel produced in Europe from the early Iron Age to the appearance of the blast furnace (1350). Small quantities of natural steel continued to be produced by bloomsmiths, especially in the bog iron furnaces of colonial New England until the late 19th century.

German steel: 0.2% carbon content or greater. Steel made from the decarburizing of cast iron in finery furnaces, as, for example, at the Saugus Ironworks after 1646. The strategy of making German steel dominated European steel production between 1400 and the advent of bulk process steel technologies, hence the term "continental method" as an alternative name for this type of steel production.

Wrought steel: 0.2 – 0.5% carbon content (cc); another name for malleable iron. Wrought steel was made from iron bar stock and was deliberately carburized during the fining process to make steel tools that are still commonplace today, such as the ubiquitous blacksmith's leg vise.

Low carbon steel: 0.2 – 0.5% carbon content (cc). Less malleable and ductile than wrought and malleable iron due to its lack of ferrosilicate, low carbon steel is harder and more durable than either and can be only slightly hardened by quenching. Some recent authors (Sherby 1995a) define low carbon steel as having 0.1% cc. Produced after 1870 as bulk process steel (e.g. by the Bessemer process), low carbon steel has all its siliceous slag content removed by oxidation. Before the advent of bulk process steel production, there was no such term as "low carbon steel." All iron that could not be hardened by quenching (< 0.5% cc) was known as "malleable" iron, more recently often referred to as "wrought" iron.

Tool steel: As with many forms of iron and steel, the term "tool steel" has multiple meanings. Tool steel has traditionally been known as steel with 0.5 – 2.0% carbon content (cc). Tool steel has the unique characteristic that it can be hardened by quenching, which then requires tempering to alleviate its brittleness. Increasing carbon content decreases the malleability of steel. If containing >1.5% carbon content, steel is not malleable, and, thus, not forgeable, at any temperature. Such steel is now called ultra high carbon steel (UHCS). Palmer, in *Tool Steel Simplified*, provides this generic description of tool steel: "Any steel that is used for the working parts of tools" (Palmer 1937, 10). The modern definition (post 1950) of tool steel is any steel containing more than 4% of one or more alloys. For a more in depth description of tool steel see *Appendix V*.

Ultra high carbon steel (UHCS): 1.5 – 2.5% carbon content (cc); a modern form of hardened steel characterized by superplasticity at high temperatures and used in industrial applications, such as jet engine turbine manufacturing, where extreme strength, durability, and exact alloy content are necessary. Powdered metallurgy technology is frequently used to make UHCS.

Cast iron: 2.0 – 4.5% carbon content (cc); hard and brittle; not machinable unless annealed to produce malleable cast iron. For a more detailed description of the many varieties of cast iron see cast iron.

*Caveats to carbon content of ferrous metals

- Both modern and antiquarian sources vary widely in their definitions of wrought iron, malleable iron, and steel. Modern sources variously define steel and/or low carbon steel as iron having a carbon content greater than 0.08%, 0.1%, 0.2%, and 0.3%.

- Before the advent of bulk process steel industries (1870), which produced huge quantities of low carbon steel that could have a carbon content in the range of 0.08 – 0.5%, iron having a carbon content of < 0.5% cc was called malleable iron. Other generic terms for iron that could not be hardened by quenching (> 0.5% cc) were bar iron, wrought iron, and merchant bar.

- The 1911 edition of the *Encyclopedia Britannica* defines wrought iron as containing less than 0.3% carbon, cast iron as having 2.2% or more carbon content and steel as having an intermediate carbon content > 0.3% and < 2.2%.

- Gordon (1996) defines steel as having a carbon content > 0.2%. This cutoff point is probably the most appropriate to use in defining steel, but also poses a problem since most sources define wrought iron as having < 0.08% cc; therefore, leading to the confusion of iron with a carbon content > 0.08% but < 0.2% as being either wrought iron, low carbon steel or an orphan form of undefined iron.

- In view of the long tradition of the use of the term malleable iron, this glossary resurrects the use of that term to cover this gray area of the carbon content of ferrous metals.

Carbon content (cc): The percentage of carbon in an object. Historically, carbon content was determined by a visual examination of the visible microstructure of a broken sample of iron or steel. Modern techniques include examination of a broken sample of steel cast in a mold using rapid spectrographic analysis.

Carbon cycle: Cycle in which living organisms incorporate carbon from the carbon dioxide in air or dissolved in water through the process of photosynthesis. Carbon combined with oxygen (CO_2) is released by respiration, decay, and oxidation, as with the burning of biomass fuels. Human activity, including the smelting of metal and the use of the internal combustion engine, is rapidly increasing the amount of free CO_2 in the biosphere.

Carbon diffusion: Diffusion that occurs mechanically as a result of hammering or thermal treatments, such as annealing, but more efficiently initiated as a chemical reaction during crucible steel production. The gaseous diffusion of carbon into wrought iron, malleable iron, and low carbon steel cannot occur at a temperature less than 850° C. See carburize and case hardening.

Carbon dioxide (CO_2): Gas formed as a component of the carbon cycle when, after carbon is incorporated in living matter by photosynthesis, it combines with oxygen and is released by respiration, decay, or combustion. As a key component of the carbon cycle, carbon dioxide is produced by all biotic organisms during respiration and by the combustion of biomass fuels as occurs during the smelting of iron ore to produce iron. Carbon dioxide is also produced during the smelting process when carbon monoxide comes into contact with air and burns with a blue flame ($2CO + O_2 = CO_2$). It is reabsorbed by plants (trees, crops, grasses) during the photosynthesis process, which produces oxygen as a waste product. As a gas, the 2007 average global concentration of carbon dioxide was 383 ppm by volume in the earth's atmosphere and the rate of increase was 2.14 ppm per year. By 2012, the concentrations had risen to 394.25 ppm (http://www.esrl.noaa.gov/gmd/ccgg/trends/). Human activities are now increasing greenhouse gas levels (water vapor, CO_2, methane, ozone, chlorofluorocarbons, etc.) more quickly than in the past. As a result of this increase in the atmospheric level of greenhouse gases, more long wave radiation is being reflected downward to the earth's surface from these gases, resulting in rising worldwide temperatures and global climate change. See carbon, carbon monoxide, cataclysmic climate change, and biomass.

Carbon elimination: The decarburization of ferrous metals, such as the fining of malleable iron to make wrought iron, or pig iron to make German steel. Carbon elimination is the result of the oxidation of carbon to carbon monoxide (CO) and carbon dioxide (CO_2).

Carbon family structures: Bain (1939) classifies three principal families of carbon steel structures: 1) pearlitic lamellar structures formed by the slow cooling of austenite below a temperature of 1000° F, 2) martensitic structures characterized by spheroidal structures formed by rapid cooling at 200° F, requiring tempering for hardness mitigation, and 3) bainite structures formed at an intermediate range of transformation temperatures (800° F - 200° F) and having different mechanical properties than lamellar pearlite or tempered martensite, i.e. enhanced toughness for the same hardness as tempered martensite. During the bainite formation phase, austenite does not disappear entirely and may be retained during constant temperature transformation up to 25% of its original content. See annealing, austenite, bainite, martensite, pearlite, and transformation kinetics.

Carbon footprint: "Measure of the impact human activities have on the environment in terms of the amount of green house gases produced, measured in units of carbon dioxide" (Wikipedia 2008).

Carbon gradation: Pattern of carbon distribution that results from the cementation or case hardening processes. The heating of the iron bar stock in the cementation furnace or of tools packed for case hardening gradually moves carbon atoms from the surface to the interior, resulting in gradation patterns expressed as changes in the crystal structure. Natural steel produced in direct process smelting furnaces lacks carbon gradations due to the erratic and heterogeneous distribution of carbon during the smelting process. See cementation furnace, blister steel, natural steel, and case hardening.

Carbon monoxide (CO): Gas formed from the combustion of oxygen and carbon in the reduction of iron ore in the direct process bloomery, which, when burned in the smelting (reduction) process, also produces CO_2. Hot carbon monoxide serves as the reducing agent (electron donor) in the three-stage removal of oxygen from ferric oxide, first creating Fe_3O_4 then FeO. In the final stage, carbon monoxide removes the last oxygen molecule from the ferric oxide (FeO), creating pure iron (ferrite) and oxygen (O_2). In addition to CO_2 being emitted during the combustion of furnace fuels, emissions also occur when carbon monoxide is brought into contact with air and burns with a blue flame, producing carbon dioxide ($2CO + O_2 = 2CO_2$).

Carbon movement: "The average carbon atom moves 0.06 inches in about eight hours during carburizing" (Shrager 1949, 176). This slow movement makes the carburizing of edge tools by forge welding a very tedious process. All blacksmiths know it is much easier to make a tool out of low carbon steel bar stock than to carburize wrought iron into a low carbon steel tool.

Carbon nanotube: Cylinder-shaped graphitic layers with diameters measured in nanometers but with lengths up to a millimeter. Discovered in 1996 as a product of fullerene research by 1996 Nobel Prize laureates Robert F. Curl, Harold W. Kroto, and Richard E. Smalley, carbon nanotubes allow the production of a wide variety of sophisticated products ranging from highly flexible tennis rackets to bullet-proof clothing, sophisticated microtransistors, pharmaceutical drug delivery systems, computer and cell phone applications, and production of sophisticated superplastic metals. A major issue with carbon nanotube production is the wide variety of toxic off-gasses and other contaminants that are associated with the cultivation of carbon nanotubes. Implicit in the anticipated huge increase in carbon nanotube applications and production is the concomitant increase in volatile persistent organic pollutants, which accompany the evolution of this new industry. While the discovery of the molecular structure and potential technological usefulness of carbon nanotubes dates to the mid-1990s, the actual inadvertent usefulness of carbon nanotubes dates to the production of the Damascus sword. See graphene, nanotube, nanowires, and superplasticity.

Carbon and slag solubility: The tendency of carbon and slag to become more soluble with increasing temperature. In the smelting process, carbon and slag solubility are temperature-dependent and increase with increasing temperature. The higher temperatures achieved by blast furnaces (after 1350) were the critical factor in facilitating carbon and slag solubility and, thus, the efficient production of cast iron.

Carbon steel: If no alloys are added, all steel is a solid solution of iron (ferrite) and carbon. Pure low carbon steel (> 0.5% carbon content) is a form of malleable iron with no slag impurities, the mass production of which was made possible by the implementation of the Bessemer pneumatic process and other bulk steel production methods that oxidized all slag constituents. Due to its lack of siliceous slag, carbon steel is not corrosion resistant, nor is it as malleable or ductile as wrought iron, but more malleable than tool steel. Low carbon steel is very useful for manufacturing drop-forged tools, especially if mixed with alloys, but not appropriate for the manufacture of edge tools. The low carbon steel produced by the Bessemer process had limited useful applications until manganese-containing spiegeleisen was added in the smelting process. Carbon steel produced by early direct and indirect process as well as in most other modern steel production processes contain varying amounts of naturally-occurring manganese < 1.65%, silicon < 0.90%, phosphorus < 0.15%, copper < 0.6%, and sulfur < 0.05%. Carbon steels not exceeding these naturally-occurring alloy limits are called "standard carbon steels". See carbon steel classification, alloy steel, Bessemer process, carbon content of ferrous metals, manganese, and spiegeleisen.

Carbon steel classifications: Modern carbon steel is now classified by the AISI into the following categories of standard steels based on the carbon content (cc) with the limitations noted above for maximum naturally-occurring alloys.

- Low carbon steel: < 0.2% cc
- Medium carbon steel: 0.2 to 0.45% cc
- High carbon steel: 0.45 to 1.5% cc
- Steel used to make most hand tools is noted as having 0.5 to 0.62% cc
- Alloy steels: carbon steels containing < 3.99% chromium or other alloys
- Stainless steels: a type of standard carbon steel containing 4% to 30% chromium. See stainless steel.
- Tool steel: not included in the category of standard carbon steels, tool steel is

defined as steels containing more than 4% of one or more of a variety of alloys. See tool steel, *Appendix V*.

Carbon steels are further classified as:

- Nonresulfurized with a manganese content < 1.0%
- Nonresulfurized with a manganese content > 1.0%
- Resulfurized free machinery carbon steel
- Resulfurized and rephosphorized free machinery carbon steel

Carbon tax: The taxation of the industrial, commercial, and petrochemical (e.g. coal, automobiles, aircraft, etc.) production of climate-altering carbon dioxide (CO_2).

Carboniferous materials: Carbon-bearing materials used in a cementation furnace to make steel as biomass fuels or as a source of carbon for the case hardening of low carbon steel tools, e.g. charcoal dust, charred bone, bone meal, wood, charcoal, coal, pigskin, hides, et al.

Carbonless steel: Modern form of steel, made without carbon using elements, such as manganese and titanium, which produce an alloy with microstructural characteristics similar to an iron-carbon alloy (steel). Wertime (1962) is an early commentator on the mid-20th century development of this oxymoronic form of steel.

Carburize: To diffuse carbon into wrought iron or steel at temperatures above 850° C – 910° C but below the melting point of the iron or steel artifact being carburized in a carbon rich environment of charcoal, coke, or coal. Carburizing sheet iron into sheet steel by submergence in a charcoal fire (case hardening) was an ancient steelmaking tradition of desert swordsmiths and medina-dwelling cutlers. Partially enclosing an iron tool within a charcoal fire for a period of 12 to 36 hours was another variation of the carburization process for making an edge tool. However, the iron shaft of the tool being carburized had to be enclosed in clay or some other covering to protect it from the oxidizing effect of the fire.

Cascading Industrial Revolutions:

- Stone Age
- Bronze Age
- Iron Age
- First Proto-Industrial Revolution (1300-1739): The age of the blast furnace, printing press, round world exploration using wooden ships built by the edge tools of ironmongers.
- First Industrial Revolution (1740-1839): Age of Coke, cast steel, reverbatory furnace; the steam engine as prime mover.
- Second Industrial Revolution (1840-1874): Age of iron, factory system of mass production, perfection of the production of malleable cast iron and grey cast iron;

the railroad as prime mover.

- Third Industrial Revolution (1875-1899): Age of steel, perfection of Bessemer's bulk process, Siemens' open heart and Mushet's alloy steel production; introduction of structural steel to build cities, use of the telegraph, telephone, and electric lights; electricity as prime mover.
- Fourth Industrial Revolution (1900-1944): Age of petrochemical-electrical man, automobile, radio, diesel engine, age of global warfare; the gasoline engine as prime mover.
- Fifth Industrial Revolution (1945-1985): Age of petrochemical-electrical-nuclear man, atomic bomb, TV, suburbia, leaded gasoline, industrial agriculture, pesticides, chemical fallout; petrochemicals as prime mover.
- Sixth Industrial Revolution (1986-2010): Age of information and communications technology, Chernobyl, microchips, microprocessors, internet, fiber optics, Windows, hyper-connectivity; digital technology as prime mover.
- Seventh Industrial Revolution (2011f.): Age of biocatastrophe, round world impact of cascading Industrial Revolutions, Fukushima Daiichi, Arab Spring, world water crisis, cataclysmic climate change, antibiotic resistant bacteria and viral infections, bioterrorism, ascendency of the klepto-plutocracy and its shadow banking network, declining public resources, increasing public needs, political paralysis, growing income disparity, assault rifle as an icon of human freedom; shadow banking network as prime mover.

Case carburization: Ancient, as well as modern, toolmaking technique of carburizing tools in charcoal pits, ideally at a temperature at or above 850° C - 910° C, where the increased solubility of carbon results in the production of layers of ferrite and cementite (pearlite). The various crystalline patterns of pearlite that result are dependent on the relative abundance and grain size of microstructural formations of ferrite and cementite created during the carburizing process, which, in turn, are dependent on carbon content, cooling rates, additional thermal treatments, and tempering. See carburize and case hardening.

Case depth: Depth of carburization, which is time, temperature, and carbon content dependent.

Case hardening: Process of creating a hardened and long-wearing thin outer edge of steel on forge welded tools. The inner component of the tool remained soft and ductile. During case hardening, selected tools were placed in a protective enclosure with carboniferous material and heated to harden the outside surface. The development of the cementation furnace (after 1650), which protected the enclosed wrought iron from the oxidizing effects of the fire, was simply case hardening on a grand scale. The interior bundles of wrought iron in the cementation furnace always had a lower carbon content than the hardened exterior bundles. In the 19th century, case hardening became widely utilized on malleable cast iron and drop-forged carbon steel tools after they had been shaped and machined by annealing or slowly reheating and cooling the tool. The notation "case hardened" is often found on L. S. Starrett and other precision tools, such as combination squares. See carburize and case carburization.

Cassiterite: Alluvial tin, found in Cornwall, England, Thailand, Malaysia, and desert areas of Egypt. It was essential for the production of tin-bronze tools, which replaced arsenic-bronze tools in the early Bronze Age. Interruption of tin trade routes may have played a role in the rapid substitution of iron for bronze in the late second millennium BC (+/- 1200 BC), first in Cyprus and then in Crete and the Peloponnesus region of Greece.

Cast iron: Carbide of iron, i.e. iron with such a high carbon content (2.0% - 5.0%) that it is not malleable at any temperature. Cast iron was originally a waste product of early shaft furnaces that ran too hot and, later, the principal product of blast furnaces. Traditionally cast into pigs before being remelted in special function furnaces, cast iron was also often cast directly into cannons and hollowware prior to the era of bulk-processed steel produced from liquid cast iron. After the widespread appearance of the cupola furnace, most cast iron was reheated and refined to further remove impurities and excess carbon before being cast into specific forms, e.g. wheel axles, agricultural equipment, boilers, etc. It was also reworked as wrought or malleable iron, carbon steel, or crucible steel. According to Edward Knight, "The new production, instead of remaining a spongy mass in the midst of the coal and slag, to be hauled thence by tongs and be subjected to the hammer to clear it of foreign matter and condense it, ran down like a fluid to the bottom hearth, and found its way out of a hole, running into a form according to the shape of the hollow in the ground which received it. By this means the smelter found he could withdraw the iron, without tearing out the front of his furnace for each bloom" (Knight 1877, 1378). Knight's Victorian era description of the production of cast iron as a new innovation, even though it was not, is evidence of the continuing widespread use of direct process bloomery smelting in mid-19th century American iron production facilities. See malleable cast iron, white cast iron, and grey cast iron.

Cast irons: Cast iron comes in many forms, the following of which are the most important:

	Silicon %	Graphite % (Free Carbon)	Combined Carbon %	Total Carbon %	Properties
White cast iron	0.70	0.10	2.65	2.75	Very hard
Annealed malleable iron	0.70	2.70	0.05	2.75	Machinable
Cast iron for chilled castings	1.00	1.00	2.00	3.00	Very hard
Semi-steel	1.75	2.80	0.40	3.20	Machinable
Grey cast iron	2.00	3.10	0.30	3.40	Machinable

	Silicon %	Graphite % (Free Carbon)	Combined Carbon %	Total Carbon %	Properties
Soft grey cast iron	2.50	3.50	0.15	3.45	Machinable

Table 2. (Spring 1917, 180)

Modern *Machinery's Handbook* (Oberg [1914] 1996) defines ductile cast iron as characterized by graphite in ball-like forms instead of in flakes as is in grey cast iron, and as having high castability, machinability, and corrosion resistance. This is the contemporary term for malleable cast iron, which Spring (1917) notes as annealed malleable iron.

Cast steel: 1) Term specifically referring to the steel used in the late 18th and 19th century for the forging of edge tools for shipwrights, woodworkers, and other trades (e.g. scythe blades, watch springs). [Not to be confused with the more modern use of the term with respect to Bessemer's single process steel production (1865).] Listed by Moxon ([1703] 1989) in 1677 as one of the principal forms of steel available to blacksmiths at that time, Smith (1960, 24) notes that Robert Hooke's 1675 diary also references cast steel, suggesting that "cast steel was commonly known in England long before the time of Robert (sic Benjamin) Huntsman." Both writers were almost certainly referring to Wootz steel imported from Persia. Cast steel became better known after Benjamin Huntsman adapted the crucible steel process (1742) to make pure steel for his watch spring business. Edge toolmakers soon recognized the higher quality of crucible steel and marked their edge tools "cast steel" to advertise the superiority of the products over the less homogeneous blister steel and its derivatives, shear and spring steel, previously used for edge tool production, especially as weld steel. The words "cast steel" continued to be stamped on English and American made edge tools into the early years of the 20th century. The use of the electric furnace and the introduction of a wide variety of high speed and special purpose alloy tool steels beginning in the early 20th century coincided with the decline and end of crucible steel production. After 1930, the one piece cast steel edge tools produced by Collins and other companies were not made from crucible cast steel, but from steel produced in the electric arc furnace. **2)** (Modern definition) Term referring to steel made in modern bulk process furnaces, where the steel is poured directly into molds that create the shape of the final product being manufactured. Cast steel made in crucibles for the production of edge tools (see 1.) was made in batches of 10 to 50 kg; modern "cast steel" is made in batches that can range up to 50 tons, usually in the form of "low carbon steel" (0.2 – 0.5% carbon content). Most bulk process modern steel, if not rolled into sheets, is cast into one form or another. See Bessemer process, crucible steel, furnace types, ingot, and Siemens-Martin process.

Cast steel microstructure: Characterized by coarse grained austenite with coarse pearlitic inclusions and a uniform carbon content due to slow cooling in crucibles.

Casting: The pouring of liquid metals into molds to achieve a particular shape or form;

also, the shape thus obtained.

Cataclysmic climate change: Climate change characterized by unanticipated accelerating rates of global warming and sea level change due to the synergistic effects of increasing greenhouse gas emissions, methane releases from permafrost melting, decrease in albedo from melting polar ice, deforestation caused by urban and suburban development and biofuels production, and rapid melting of glaciers due to increasing temperatures and water-flow-induced slippage.

Catalan furnace: Bowl type "low bloomery" direct process furnace developed in Catalonia (Spain) after 1200 AD and still utilized as recently as the early 20th century for the production of wrought and malleable iron; a precursor of the more efficient American bloomery furnace. Depending on the fuel to ore ratio, open-hearth Catalan furnaces operated by knowledgeable forge masters could be and were used to produce small quantities of raw steel, which could be reforged into edge tools and weapons. See natural steel and Biscayne ax.

Caulking: The process of inserting oakum between ship planking by the use of caulking irons and caulking mallets to prevent leakage. During ship repair, the seam would first be opened by a reefing hook to remove old oakum. A clearing iron, also called a reefing iron, was used for further cleaning of the seam prior to re-caulking. Common caulking tools include: the boot, spike, butt, making (common caulking), crooked (bent), and deck irons (Sellens 1990), as well as special purpose long-handled horsing irons. Before 1800, caulking irons were a common product of the shipsmith. In North American colonial era shipyards, hemp and pitch-soaked cotton were sometimes used instead of oakum (Story 1995). America's most famous producer of cast steel caulking irons was Christopher Prince Drew of Kingston, Massachusetts (1815 - 1907). Numerous examples of caulking tools and mallets are on exhibit at the Davistown Museum in Liberty, Maine.

Cavitation: Formation of voids and cavities within ferrous metal microstructures, which inhibit plasticity by interfering with the uniform distribution of carbon in high quality tool steels.

Celtic metallurgy: Knowledge of ferrous metallurgy, including a wide variety of steel- and toolmaking strategies and techniques brought to central Europe by migrating Celtic communities moving westward through Europe from the western steppes of Russia. Celtic metallurgists were responsible for the flowering of the early Iron Age at Halstadt and La Téne, respectively, beginning about 800 BC. The source of their knowledge of ferrous metallurgy is unknown, but they were responsible for forging the swords, knives, and edge tools in ancient Noricum that first led to the success of the Roman Republic and the Roman Empire and then led to the defeat of the Roman Empire at the battle of Adrianople (400 AD) by Goths supplied with Celtic-made, doubled-edged, pattern-welded broad swords. Celtic metallurgists were already working in Britain when Caesar invaded in 54 AD. See Halstadt, La Téne, gladius, spathas, natural steel, and manganese.

Celt: Earliest form of stone tool that was attached to a handle of antler or wood, often secured by leather thongs.

Cementation: The process of introducing elements, including carbon, into the outer layer of a metal (e.g. iron) by means of high temperature diffusion. See case hardening.

Cementation furnace: Enclosed stone or firebrick box containing alternate layers of a carboniferous material, such as charcoal dust and the iron bar stock to be converted into blister steel. With the fire located in an open-hearth below the box, which keeps the fuel separate from the iron bar stock being carburized, the airtight furnace is fired for periods ranging from 8 to 12 days and has a capacity up to 10 tons. First noted in 1574 (Wayman 2000), its first documented use was in Nuremberg in 1607; its first documented use in England was in 1686. The use of the cementation furnace to make blister steel was the principal means of producing steel in Britain (1700-1850). Less widely adopted in continental Europe, it was supplemented by the crucible steel process after 1750 and was gradually replaced by modern steelmaking technologies after 1865. Blister steel is used for both shear (sheaf) steel and crucible steel production. The cementation steel furnace represents an industrial scale application of the case hardening principle. Steel bar stock with a widely varying carbon content could be produced depending on exposure time to carboniferous materials, such as charcoal dust, in the cementation furnace. Steel with a more uniform carbon content could be produced with lengthy firings (8-14 days). Blister steel was often piled and welded into special steels, such as shear steel for cutlers and edge tools, before cast steel became widely available (1750-1775). See blister steel, shear steel, spring steel, cast steel, and converting furnace.

Cementation steel: Steel created by heating wrought iron in powdered charcoal in a cementation furnace, which keeps the fuel separate from the iron bar stock being carburized; also called blister steel. Because of the long cooling time for most blister steel, ferrite-cementite matrices were its predominant microstructural constituents.

> Sheffield, England steel makers have been very successful in the manufacture of cementation steel. Their usual method is to pack flat strips of the best Swedish Walloon iron in charcoal inside rectangular stone boxes about four feet wide, three feet high and fourteen feet long. Alternate layers of small-sized charcoal and thin iron bars are piled in these boxes until they are filled, the bars not being allowed to touch one another. When full, top slabs are luted on to the boxes to make them airtight. Fire is kindled in the firebox below and the heat gradually raised until furnace and boxes are cherry-red in color. This heat is maintained for seven to eleven or more days, depending upon the hardness desired, i.e., the amount of carbon they desired absorbed. The furnace is closed and allowed to cool slowly, which requires another seven or more days. Upon unpacking the furnace, the bars are found to be brittle and of a steely fracture instead of the soft malleable material that was put in; they have become high carbon steel. Expert workmen are able to judge very closely the hardness of the steel by looking at the fracture and they sort the bars in this way, piling bars of similar hardness together. Bars thus made show many blisters on the surface and the steel became known as "blister steel" on this account. (Spring 1917, 116-117)

See blister steel and cementation furnace.

Cementation temperature: Temperature required for the efficient diffusion of carbon during the cementation process, from 1050° to 1100° C.

Cementite: "Iron carbide (Fe^3C) in the microstructure of steel and cast iron" (Gordon 1996, 308). Cementite is the most common iron carbide and is hard and brittle. The carbon content of cementite is all in the chemically combined form, with no free graphite flakes. The carbon content of pure cementite is 6.67% (Pollack 1977). Cementite is often fully dissolved in austenite with proper heat treatment, or, otherwise, partially dissolved in steel and iron in various grain sizes and patterns, giving mechanical properties to steel and iron products (strength, hardness, brittleness, etc). As carbon steel, cementite forms directly from the melt in the case of white cast iron. As a component of carbon steel, it either forms from austenite during cooling or from martensite during tempering. It mixes with ferrite, the other product of the cooling of austenite, to form lamellar structures called pearlite and bainite. Uniformly dispersed particles of cementite, usually with an acicular microstructure, are an essential component of martensite. The uniform homogeneous distribution of small cementite spheroids differentiates cast steel from other forms of steel, such as blister and natural steel, which are characterized by a more heterogeneous distribution of cementite. See austenite, cast steel, lamellar family of structures, martensite, pearlite, and white cast iron.

Cementite plates: Characteristic of the microstructure of steel with a carbon content of 1.0 to 2.0%. Cementite plates result in high hardness and low ductility requiring reprocessing by tempering to produce fine grained (usually spheroidal) microstructures, sometimes in the form of nanotubes and nanowires, that are no longer brittle, i.e. their superplasticity at high temperatures is demonstrated by plasticity and a lack of brittleness at room temperature. See cementite, nanotubes, nanowires, and superplasticity.

Chafery: Forge for the further reworking of anconies of wrought or malleable iron by re-tempering and reshaping or slitting the metal into forms compatible with the production of specific tools and iron products, e.g. nails, rolled sheet iron, wrought iron hardware, and malleable iron agricultural tools. See ancony, bloomery, direct process, and muck bar.

Chalybean or **Chalibean steel**: High nickel content steel smelted from the self-fluxing iron sands of the Black Sea c. 1900 BC, at the height of the Bronze Age; often mistaken for meteorite-derived steel (Piaskowski 1982); described in Greek literature by Aeschylus and Strabo. The use of iron sands by Chalybean bloomsmiths or samurai swordsmiths as the highest quality special purpose iron ore available was a prelude to the late 20th century development of powdered metallurgy, the essential ingredient of which is a very fine grained iron oxide powder. See samurai sword and Tamahagane.

Charcoal: The principal fuel used to make wrought iron and cast iron from the early Iron Age until the mid-18th century, when coke became available as a blast furnace fuel in England. Charcoal continued as the main source of fuel in heavily wooded environments, such as the US and Sweden, for a century or longer than in forest-depleted England. The very low sulfur content of charcoal iron made it ideal for edge tool production, in contrast to coke and coal fired blast furnace iron, which was high in sulfur and ash. Swedish charcoal iron, also low in phosphorous, was the key ingredient for blister steel production in converting furnaces and was then used by Sheffield crucible steel manufacturers for the final smelting of cast steel. Swedish charcoal iron was also imported in large quantities to New Bedford (and many other U.S. ports), 1815 - 1850, presumably for use by New Bedford whale crafters and edge toolmakers (Lytle 1984). Charcoal wrought iron was produced by American direct process bloomeries for the Pittsburgh crucible steel industry from its inception in 1860 until the end of the 19th century.

Charcoal hearth iron: Iron produced using charcoal as a fuel; also called charcoaled wrought iron.

Charcoal iron: Highly refined variation of iron made by alternating layers of pig iron and charcoal in a finery furnace. After being partially melted, most of the silicon and part of the phosphorus was eliminated. After solidification, the mass of iron was broken up and reheated in a second furnace with alternate layers of charcoal, which removed most of the remaining carbon, but some slag remained entrained within the bloom. Charcoal iron produced by this method contained a higher carbon content than wrought iron but contained less slag, most of which was in the form of iron silicate. Charcoal iron is, thus, a form of malleable iron containing 1% or less of silicon slag content yet is more ductile than wrought iron due to the extensive additional mechanical treatment to which it has been subjected (e.g. rolling), especially in the direction opposite from its fibrous slag layers (Aston 1939). Some Swedish bar iron produced in the late 18th and 19th centuries may have been refined by this alternative method of iron production and utilized for edge tool production via cementation furnaces. See bloom, cementation steel, malleable iron, and slag.

Chasing: Sinking of ornamental designs onto the surface of malleable sheet metals by using a jeweler's hammer and cast steel chasing tools, resulting in the formation of grooves, indentations, and furrows in the metal surface; used especially in jewelry design and manufacture. See repoussé.

Chemistry of carburization: Absorption of carbon by iron and steel from carbon monoxide, which is formed by the reaction of the oxygen in the enclosed environment of, for example, the cementation furnace, with the carbonaceous material inserted with the iron bar stock. Iron carbide (cementite) and carbon dioxide are the resultant products. The carbon dioxide is reconverted to carbon monoxide by reacting with more carbon to extend the cycle of carbide formation:

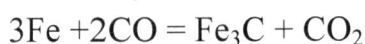

$$3Fe + 2CO = Fe_3C + CO_2$$

$$CO_2 + C = 2CO$$

See carbide.

Chill: To use cold metal inserts within mold cavities in the production of cast iron artifacts to facilitate rapid cooling and produce castings with hard surfaces but soft interiors.

Chilled cast iron: Cast iron that retains carbon in its chemically combined form if insufficient cooling time is allowed for the precipitation of graphite flakes, as occurs when the iron is cooled rapidly in iron molds rather than slowly in sand molds. "Cast iron with a low silicon content cooled rapidly enough to form cementite rather than graphite in its microstructure. Molders placed chills in molds to get white iron at places in a casting where extra hardness was desired, as on the rim of a railroad-car wheel" (Gordon 1996, 308). The characteristics of chilled cast iron are a function of the thickness of the chill (iron mold) and the carbon, silicon, sulfur, and other alloy content of the molten cast iron. See white cast iron and combined carbon.

Chinese steel: Steel made by the Chinese utilizing multiple methods including: immersing bars of wrought iron in a crucible containing melted cast iron (similar to Brescian steel production); making steel from ". . . charcoal refining of pig iron, probably a process not dissimilar to the Styrian process 1500 years later" (Barraclough 1984a, 29-33); and malleableizing white cast iron and steeling the edges of woodworking tools made from the cast. Barraclough (1984a) describes an adz c. 500 BC as follows: "The fascinating feature of this implement is that the cutting edge has been sufficiently decarburized to produce a steel-like structure; this is quite uniform and would appear to be intentional . . . The methods of 'malleableizing' cast iron castings were obviously well understood at this period since agricultural implements cast to shape and then fully malleableized were fairly widespread during the Menchius era of the 4th century BC . . . this was the process rediscovered by Réaumur some 2000 years later; in China it had gone out of use in about the 7th century AD, after 1000 years of use." Barraclough (1984a) also notes several modifications to the Brescian style co-fusion process, including a solid diffusion method, whereby coils of wrought iron were entwined with insertions of cast iron to produce lump or raw steel. See Needham (1958), malleable cast iron, white cast iron, and Brescian steel.

Chloanthite: Nickel arsenide compound composed of iron, nickel, cobalt, arsenic, and sulfur added to the iron sand smelting process to produce the famed nickel steel made by the Chalybeans of northern Anatolia, c. 1900 BC, at the height of the Bronze Age, a process noted by Aeschylus and Strabo. See Amisenian, Chalybean, iron sand smelting processes, Wertime (1980), and Piaskowski (1982).

Chrome steel: High grade alloy tool steel produced in the electric furnace and used for applications requiring great hardness, such as ball bearings. Variations include chrome vanadium steel used in automobile frame and knife and tool production and nickel chromium steel used for automobile gears, axles, armor plates, projectiles, and airplane parts. See alloy, alloy steel, stainless steel, and tool steel.

Chromium: Key alloy in the smelting of stainless steel. Chromium "is a strong carbide former and stabilizer when present in small amounts (~ 0.5%). Chromium increases hardenability, corrosion, heat, and oxidation resistance; improves high temperature strength (in conjunction with Mo)… Chromium carbides are hard and wear resistant and *increase edge holding quality* [italics added]" (Sinha 2003, 1.23). Chromium is often present in alloy steels (> 4%), tool steels (4-10%), and stainless steels (>10%). See alloy steel, stainless steel, and tool steel.

Cire Perdu: Lost wax method of making castings by using wax to make a pattern of the object being cast. After being set in plaster, the wax is melted out of the mold; the resulting mold often has great detail. This method was used for casting bronze objects in antiquity and has been recently revived by contemporary artists. See embossing sequence and drag.

Classification system for steel: The American Iron and Steel Institute (AISI) and the Society for Automotive Engineers (SAE) have developed a system for designating carbon and alloy steels. The AISI has included additional classifications for tool steels with alloy contents above 4% and stainless steels with chromium content above 10%. This system begins with a designation of 10xx for carbon steels, the xx being the carbon content. Eleven classifications of alloy steels include 66 types; AISI tool steels include 64 types. Stainless steels include 4 classifications. See *Appendix IV* for the AISI-SAE system of designating carbon and alloy steels.

Clay ironstone: Common form of iron ore and the principal form of iron ore in the Weald (Sussex), England's most important iron-producing region of the 16th century. It occurred in the form of siderite as a hard gray rock within the clays, silts, and sandstones of the Wealdean beds. This ironstone contained magnesium carbonate ($MgCO_3$) and manganese carbonate ($MnCO_3$), which, as with the manganese in the Austrian ores of Noricum, played an important role as a flux, enhancing carbon uptake in the production of raw steel (Cleere 1985). See natural steel, Weald of Sussex, and manganese.

Cleavage plane: Smooth fracture occurring parallel to one or more crystallographic planes. See deformation and lattice structure.

Close-packed hexagonal structure (CPH): Lattice structure characteristic of metals that lack plasticity or lose plasticity during cold forming, including antimony, beryllium, cadmium, cobalt, magnesium, titanium, and zinc (Shrager, 1949). See body-centered cubic unit and face-centered cubic unit.

CNC technology: Computer numerical calculation (CNC) technology references the ongoing tsunami of computerized manufacturing processes and industrial product management using computer assisted design (CAD) programs. These programs, the progeny of the information technology revolution, have vastly increased the efficiency and productivity of both small and large corporations of every description, including our rapidly expanding shadow banking network (SBN) and its numerous parasitic hedge fund and derivative marketing entities. Its downside is the concomitant growth in structural unemployment whereby large components of the middle class, formerly employed in manufacturing or now computerized service industries, become members of an underclass of unemployed, underemployed, or obsolete workers. The techno-elite who make and operate the computers that drive the Age of Information Technology represent a tiny percentage of a workforce whose main function is to serve as consumers in a now globalized market economy. Invisible underground communities of self sufficient artisans and craftspersons, including ironmongers, will inevitably utilize and benefit from CNC technologies despite the gross income inequality generated by a global klepto-plutocracy. The complex electrical power grid that is the prime mover of CNC technologies is particularly vulnerable to geomagnetic (solar) storms. Recent Homeland Security, NRC, and Oak Ridge National Laboratory reports postulate possible geomagnetic induced current (GIC) disruptions of US electrical power grids ranging up to two years or more (http://www.davistownmuseum.org/solarflares.html). See geomagnetic induced current, geomagnetic storms, klepto-plutocracy, and shadow banking network.

Coal: Mineral fuel with two common forms: 1) bituminous (with a long flame and containing 70 - 90% carbon) and 2) anthracite (with a short flame but containing 90 - 95% carbon.) Both forms contain significant quantities of mercuric sulfide, which is released to the environment during combustion in the form of particulates subject to tropospheric transport mechanisms, followed by conversion to toxic methylmercury by bacterial action. Coal was used as fuel in blast and reverbatory (puddling) furnaces but was unsuitable for bloomery and finery furnaces due to high sulfur content (+/- 2.0%). See methylmercury, biocatastrophe, and biomass fuels.

Coalbrookdale: Important 17th and 18th century iron-producing center located on the headwaters of the Severn River, in the Midlands of England, just west of Birmingham. Coalbrookdale was the location of Abraham Darby's first coke-fired blast furnace and also the location of the famous Ironbridge, England's first cast iron bridge. It may also have been the site of one of England's first cementation furnaces, constructed by Sir Basil Brooke (1636?) (Barraclough 1984a).

Cobalt: Important alloy in modern steel production. Cobalt decreases hardenability when alloyed with other elements but increases red hardness when alloyed with vanadium, chromium, or tungsten. In the forging of the Wootz steel Damascus sword, cobalt as a micro-constituent of carbide nanotubes, especially if in the presence of chromium as another micro-constituent, increases the distortion of unit cell patterns, thereby increasing strain hardening and helping to give the Damascus sword its famed hardness and cutting ability. Cobalt is often used in modern high speed drill bits. See alloy, Damascus sword, nanotubes, nanowires, red hardness, and Wootz steel.

Co-fusion: Variation of the Brescian process for producing steel, whereby wrought iron was dissolved in molten cast iron. One of the most ancient steel producing technologies, co-fusion was a variant of crucible steel production methods used in India and Asia, was later patented in England by Vickers (1839), and thousands of tons were produced annually (Barraclough 1984). Similar combinations of ore were fired in the first Siemens-

Martin open-hearth furnace (+/- 1870). See Brescian steel and Chinese steel.

Cogging: Hammering, rolling, or pressing a red-hot tool steel ingot to convert it to a billet.

Coke: "Porous, cohesive carbon made by expelling the volatile constituents from bituminous coal" (Gordon 1996, 308). A product of the partial combustion of coal, usually in large brick ovens, coke was invented by Abraham Darby in England in 1709, and first produced commercially in 1735. Its use in blast furnaces resulted in the production of iron with a high sulfur content, requiring a calcium bearing flux (lime) for slag removal, as well as higher firing temperatures. The availability of coke as a fuel stimulated the invention of the steam-driven piston bellows, which was needed to provide a more powerful air blast for the smelting process. After 1785, the efficient use of coke stimulated production of larger quantities of higher quality cast iron for reprocessing as wrought and malleable iron in the newly invented reverbatory puddling furnace. See reverbatory furnace and puddling furnace.

Coke smelting: Use of coke, instead of charcoal, as a fuel to smelt iron; became widely used in England after the middle of the 18th century due to the depletion of English forests by charcoal production. Coke smelting didn't become widespread in the United States until the 19th century due to the extensive availability of wood from which to make charcoal fuel.

Cold blast/hot blast: The switch from using a cold air blast to a hot air blast after Nielson's 1828 innovation of using waste gases to preheat the blast greatly increased the efficiency of 19th century English, American, and European blast furnaces.

Cold drawing: Reshaping of iron or steel bar stock by drawing it through dies, resulting in its permanent deformation.

Cold forging: Ancient tradition of hammering cold, malleable iron into tools and armor. Bloomery-smelted charcoal iron with slag impurities, especially silicon, lent itself to toolmaking because it welds easily, does not split, and is malleable to mold with a hammer and anvil even at low temperatures. Modern low carbon steel made in blast furnaces with mineral fuel, gas, or electricity lacks the silicon inclusions that facilitated the creative cold hand-forging of Swedish and American charcoal irons (Bealer 1976, 36). See charcoal iron.

Cold short: Wrought iron containing phosphorous, thus brittle if worked or hammered at room temperature.

Cold working: Rolling or forging steel at temperatures below 400° C to avoid re-crystallization. Cold working increases strength and decreases ductility and toughness of steel. If high in phosphorus, cold worked wrought iron can reach a hardness level as great as 300 HV, above that of non-heat-treated carburized steel (Shrager 1961). See phosphorous and carburize.

Collier: 1) Producer of charcoal for bloomeries, fineries, and charcoal-fired blast furnaces. The activity of collier resulted in the deforestation of large areas of Europe and the Mediterranean before 1800, and eastern North America until the Civil War. **2)** Coal miner. **3)** A ship that transports coal.

Combined carbon: Carbon chemically combined in iron and steel alloys and, therefore, not existing as free graphite flakes. As contained in rapidly cooled white cast iron, the transformation of combined carbon into free carbon is the key step in the successful production of malleable cast iron, which is thus more machinable. See malleable cast iron, and white cast iron.

Combustion gasses: Gasses produced in smelting and fining furnaces and forges that oxidize carbon and prevent carburization. The most important combustion gas is carbon monoxide (CO), which is formed, in part, during combustion due to lack of oxygen; otherwise, nonflammable CO_2 (carbon dioxide) would be produced. During the reduction process, if sufficient oxygen becomes available, CO will burn with a blue flame, producing CO_2. CO as a fuel is the most significant combustion gas due to its ability to react with steel surfaces by oxidation. Such combustion of CO produces iron oxide (FeO), which is then sometimes reused (reduced) in the smelting process to produce pure iron, illustrating the complex dynamic of combustion/reduction processes. See carbon, carbon dioxide, carbon monoxide, iron oxide, and reduction.

Compressive strength: Relative ability of a metal to avoid failure when pressed or squeezed.

Computer-aided design (CAD): Also known as computer-aided drafting, it is the use of computer systems to assist in the creation, modification, analysis, or optimization of a design. These computer programs are the prime movers of CNC technologies in the Age of Hyper-digital Technocracies.

Computer numerical calculations: See CNC technology.

Contaminant pulse: Presence of ecotoxins in abiotic and biotic media. The appearance of global contaminant pulses is a result of the advent of bulk steel production (1870), the widespread use of the internal combustion engine (> 1910), and the development of petrochemical industries producing pesticides, sophisticated arms and explosives, and globally traded consumer products and electronic equipment (> 1940). Global and regional contaminant pulses of biologically significant chemical fallout (BSCF) including persistent organic pollutants (POPs) and other ecotoxins can be tracked and monitored as contaminant signals in abiotic and biotic media, often measured in parts per billion or parts per trillion. Many forms of BSCF bioaccumulate in pathways to human consumption. Concentration factors from water to consumers at the highest trophic levels (e.g. predatory birds, humans) maybe as high as six orders of magnitude. See biocatastrophe, biologically significant chemical fallout, petrochemical-electrical man, and tragedy of the world commons.

Continental method: Generic name for steel produced by decarburizing cast iron (German steel), so called because it was the principal method of producing steel in continental Europe even after the cementation furnace dominated steel production in England (1700). See German steel.

Converting furnace: Airtight furnace, also called a cementation furnace, usually of a rectangular design and used to convert iron bar stock to blister steel after being packed in charcoal. The first converting furnaces to make steel appeared in Nuremburg in 1607 (Barraclough 1984a), in England in 1686, and in the United States by 1713. The converting furnace prevented contact between the fuel and the bar stock, allowing production of steel in larger quantities and of a more uniform quality than that made by fining cast iron lacking the manganese content of the ores used to make German steel. See blister steel, cementation furnace, and cementation steel.

Convivial tools: Instruments of manual operation used to harvest and/or to create art and artifact out of renewable resources (e.g. the woodworking hand tools of the shipwright) or to recycle, rework, or reforge industrial salvage. In the broader sense of the term, convivial tools are those which offer the pleasurable experience of using one's hands to create unique forms or execute significant repairs in comparison to the more impersonal experience of operating machinery to do work. See born-again ironmonger.

Cooling rate: Speed of cooling. Along with carbon and alloy content, the cooling rate of steel or cast iron after forging, melting, or austenizing plays a major role in determining the allotropic form of the ferrous metal being produced. See phase transformation kinetics.

Cooling rate of austenite: The cooling rate of austenite determines the microstructure of ferrite and cementite and their mode of distribution in formations, such as pearlite and bainite, after austenite is cooled below the critical temperature (723° C / 1300° F). Rapid cooling of austenite to a temperature of 220° F creates martensite. The halting of the cooling of austenite above the temperature at which martensite is formed, and its maintenance in a temperature range between 1300 and 220° F (typically +/-800° F) for variable periods of time creates the wide variety of microstructures characteristic of bainite. Differences in the alloy content of austenite provide additional bases for variations in microstructural formations, which may have specific industrial or commercial applications, such as enhancing strength, hardness, or durability. See austempering, bainite, critical temperature, critical cooling rate, and martensite.

Cope: Top half of a foundry mold. See drag and foundry.

Copper alloys: The two most important copper alloys are brass (an alloy of copper and zinc) and bronze (an alloy of copper and tin). Usually both contain small amounts of lead. See alloy.

Coppice: Wood harvested from the undergrowth of a wide variety of trees and used to make charcoal. In England, small-sized wood, including branch tops with a diameter of 6 cm. or less (2.26 in.), were used to make charcoal; larger wood sizes produced friable charcoal that transport or weight of the charge in the bloomery or blast furnace would reduce to dust. The small diameter of the wood used to make charcoal meant that regenerative coppicing renewed charcoal sources on the estates of the landed gentry, who supplied much of the furnace fuel in England before the age of coke (1709). See Weald of Sussex, charcoal, and collier.

Core box: Wooden molds used to form sand cores, which replicate holes, cavities, and pockets in foundry castings. See pattern molding and foundry.

Core maker: One who designed and constructed the wooden molds used for foundry work, along with a patternmaker. Sometimes, the core maker and the patternmaker were the same person; at larger foundries, these functions were often two separate trades. See patternmaker.

Coring: A procedure for cutting or shaping the opening in the middle of an ax or adz for hafting of the handle.

Coronal mass ejection (CME) event: The ejection by the sun of a blast of plasma composed of electrons, protons, and highly excited gas at hundreds to thousands of kilometers per second. A CME event is the source of most, but not all, geomagnetic induced currents (GIC). For more information about geomagnetic storms and their potential impact on the power grid go to http://www.davistownmuseum.org/solarflares.html. The impact of solar storms should be of interest to all contemporary metalworkers and born-again ironmongers in that the worst case scenario for electric grid failure is an outage lasting four to ten years. Non-electric powered hand tools will be an essential component of the long term response to electronic infrastructure collapse.

Corrosion resistance: Prevention of rusting provided by the addition of nickel, copper, or aluminum, as in stainless steels.

CPM S30V: "A martensitic (hardened) powder-made (sintered) wear and corrosion resistant stainless steel developed by Dick Barber of Crucible Materials Corporation in collaboration with knife-maker Chris Reeve." It is composed of carbon: 1.45%, chromium: 14.00%, vanadium: 4.00%, and molybdenum: 2.00% (Erin Casson 2013, personal communication).

Creative economy: Ability of self-sufficient, post-apocalypse, off-the-grid communities using convivial tools and renewable resources to maintain viable market, trading, and/or barter economies in a biosphere with limited biomass fuel, minerals, fresh water, and natural food resources. Ironmongers, as well as toolmakers and artisans, will be an essential component of the creative economy of post-apocalypse society. See convivial tools, biocatastrophe, and biosphere as bank account.

Creep: Slow, impermanent deformation of material under steady force that is below the elastic limit of the material being deformed, occurring over a long period of time. See deformation, elasticity.

Critical cooling rate: Lowest rate of austenite cooling at which martensite will form. The critical cooling rate is dependent on the carbon content of the steel being heat treated.

If the cooling rate is lower than the critical cooling rate, pearlite will be formed instead of martensite. "The rate of cooling which just prevents the reaction [of austenite to pearlite] in the vicinity of 1000 degrees Fahr. and results in the lower temperature reaction is called the 'critical cooling rate' or critical quenching rate. The former transformation and its temperature is spoken of as Ar' and the second as Ar'^{A}; the critical rate is the cooling rate separating the two. Many steels, however, cooled at about the critical rate, transform in part by one mode and in part by the other, even in the same grain" (Bain 1939, 29). "The critical cooling rate for hardening is decreased by an increase in austenite grain diameter" (Bain 1939, 51). See austenite, martensite, and pearlite.

Critical point: Lowest temperature at which tool steel will harden when quenched, or at which changes in the crystal structure of a metal will take place. See critical temperature, quench, and quenching threshold.

Critical temperature: Temperature (T_c, 1341° F / 723° C) at which ferrite and pearlite (alpha iron) are transformed into austenite (gamma iron). During this transformation, the microstructure of the iron crystals is changed, and the amount of carbon dissolved into this microstructure increases. As a component of heat treatment processes (e.g. sword and edge tool tempering) slow cooling will allow the austenite to return to microstructures characteristic of ferrite or pearlite. Rapid cooling changes the crystalline content of austenite to the hard and brittle microstructure of martensite and requires further tempering prior to the final forging of an edge tool or sword. See austenizing temperature, crystallography of ferrous metals, microstructure, yaki-ire, and yaki-modoshi.

Critical temperature range: Temperature difference between A_1, the eutectoid temperature, (723° C) and A_3, the alpha-iron gamma-iron transformation temperature for pure iron (910° C) expressed on contemporary phase transformation (iron-carbon) diagrams as A_{cm}. See *Appendix VI*.

Crucible: High-temperature-resistant earthenware vessel used for melting, fusing, or casting metals. Clay crucibles have been used since the early Iron Age for the smelting of small quantities of steel. The earliest documented use of crucibles for steel production was the smelting of Wootz steel in Muslim communities c. 300 BC (Sherby 1995a). Beginning in 1742, Benjamin Huntsman produced Sheffield cast steel in small clay crucibles. During the height of Sheffield crucible cast steel production in the 19th century, cast steel was produced in rows of crucibles with a +/-50 kg capacity. The lack of high temperature resistant clays delayed cast steel production in the United States until the mid-19th century, when Joseph Dixon's invention of the graphite crucible provided an alternative to Stourbridge clays for crucible production. See crucible steel, Stourbridge clay, and Wootz steel.

Crucible furnace: Furnace composed of two firebrick boxes used to manufacture cast steel for specialized purposes, such as watch springs and edge tools. The fire was in the lower box, and the upper box held the crucibles. Other forms of crucible furnaces had only a single firebox. Wasteful of heat and expensive to operate, crucible furnaces were made obsolete by the invention of the electric arc furnace. See cast steel and crucible steel.

Crucible holes: Slang term for the openings in a crucible furnace, into which the crucible itself was inserted during the production of cast steel.

Crucible steel: Steel created by melting together pieces of blister or shear steel in a clay crucible containing carboniferous material. The resulting chemical reaction equalized the uniformity of carbon distribution. In the mid-19th century, crucible steel was also produced from a combination of puddled wrought iron and steel scrap. High quality cast steel has a microstructure of small carbide spheroids and qualities of plasticity, or near superplasticity, at temperatures above its melting point, which allow it to be shaped and rolled easily. Crucible steel was the first choice of English and American toolmakers for the manufacture of high quality edge tools, often advertised by edge toolmakers as "warranted cast steel." Crucible steel was made in and imported from England after 1742 by Benjamin Huntsman and not manufactured in the United States until about 1860. Spring (1917) makes the following observation:

> For a century crucibles were made from clay molded to form, slowly dried and very carefully burned. Usually each steel maker made his own crucibles. They could be used but three times, becoming so thin and tender after use for three batches of steel that they were not safe for a fourth. Graphite crucibles are now very largely used. They withstand the severe heat much better and can be used five or six times. The expense item for either clay or graphite crucibles is a large one. After filling with small pieces of blister or shear steel the crucibles are entirely surrounded by coal or coke in the furnace pit. The fire is so regulated that the steel is not too quickly melted. Fresh coal or coke must be put in around the crucibles two or even three times. When he thinks the steel should be molten, the expert attendant known as the "melter" quickly removes the tight fitting cover of the crucible and with an iron rod determines whether any un-melted pieces remain. After complete melting the steel must be "killed," else it will boil up in the mold upon pouring and leave a spongy or insufficiently solid "ingot" or block of steel. This "killing" of steel is a rather peculiar phenomenon. It is accomplished by allowing the steel to remain quiet in the furnace for another half hour or so. Undoubtedly the quieting is the result of the escape of the gases or impurities, which are contained in the charge, and absorption of the chemical element, silicon, from the walls of the crucible. ...When the steel has been properly melted and killed it is ready to pour. An assistant lifts the cover from the melting hole; the "puller out" seizes the crucible just below the bulge with circular tongs and pulls it from the coke, which surrounds it. The slag is skimmed off the top and the steel poured into iron molds forming small "ingots," usually from 2 to 4 inches square and two feet or more long. Every part of the process, even the pouring, must be done with extreme skill and care or the product suffers. After liberation from their molds, the ingots are heated and either rolled or hammered down to the sizes desired for tools, etc. (Spring 1917, 118-120)

Crucible steel should not be confused with the "cast steel" ingots of the Bessemer process. See cast steel 1) and 2).

Crypto-fascism: Concealed agendas and hidden languages of a global consumer society that glorifies growth and consumption at the expense of stable self-sufficient communities using renewable resources in sustainable ecosystems. Crypto-fascist pyrotechnical transnational industrial society uses military policy (e.g. oil wars) to control and horde nonrenewable energy resources based on religious belief in divine right and intelligent design for privileged social classes or sects (e.g. the klepto-plutocracy, shadow banking network, Tea Party Taliban, and Christian fascists.) Crypto-fascist consumer society utilizes mass media in lieu of secret police to control social and economic activities, influence consumer habits, encourage non-sustainable growth, and marginalize opposition to the frenzied consumption of nonrenewable resources. See klepto-plutocracy, shadow banking network, and Tea Party Taliban.

Crystal: Geometric structure with a homogeneous distribution of atoms in patterns characteristic of the specific elements or compounds in that structure.

Crystal structure: As used in ferrous metallurgy, the atomic structure characterizing each phase of iron-cementite transformation.

Crystallography of ferrous metals: Study of the variations in the space lattice formations (cubic structure) of the crystal structures of iron that result from various thermal and mechanical treatments. Such structures are also dependent on the carbon and/or other alloy content of the iron. See lattice structure and space lattice.

Cubic structure: Space lattice forms of unit cells within the crystal structures of metals. In ferrous metallurgy, two forms of cubic structure characterize iron, i.e. the FCC (face-centered cubic) and the BCC (body-centered cubic). A third form of cubic space lattice is the close-packed hexagonal (CPH), characteristic of such non-ferrous metals as cobalt, magnesium, and titanium. Metals in this latter group lack plasticity (Shrager 1949, 5-6). See crystal, space lattice, and unit cell.

Cupola furnace: Nineteenth century form of blast furnace used for re-melting cast iron; a simple, fuel efficient shaft furnace usually using coke as a fuel, one of the principal means for melting cast iron for foundries. The iron is melted in contact with the fuel. Only gray cast iron is produced in the cupola furnace. Instead of the stone pyramid form of the traditional blast furnace, cupola furnaces are constructed of sheet iron shells culminating in a narrow top with adjacent elevators and a bridge to bring fuel and ore to the top of the furnace. See blast, blast furnace, cast iron, foundry, and grey cast iron.

Curie temperature: Temperature at which magnetic transformation takes place, as in the transformation of alpha iron to gamma iron (910° C).

Currency bar: Primary form of iron as a trade item in the early Iron Age, later replaced by iron bar stock after the invention of rolling equipment in the 17th century. A currency bar is a bloom of wrought iron, malleable iron, low carbon steel, or raw steel derived from a direct-process smelting furnace. A blacksmith would take this bloom, often of heterogeneous carbon content, hammer it to expel the slag and form bars of iron or raw steel of various shapes and sizes, usually with handled ends, for easy transport from smelting sites, such as those in Austria, to iron working and swordmaking centers, such as those on the lower Rhine. Metallographic analysis indicates that these bars had a widely varying carbon content, ranging from wrought iron to raw steel. Currency bars were traded throughout Europe beginning in the early Iron Age. See bloomery, Iron Road, Merovingian swords, and natural steel.

Cutler: Knife- or swordmaker. Boston swordsmiths were called "sword cutlers" c. 1800.

Cyaniding: Producing thin, exterior, high carbon layers in tool steel by applying liquid cyanide; a modern process for case hardening. See case hardening.

Cybernetics: The science of communication and automatic control systems.

Cyberterrorism: The use of internet-based attacks by cyber-terrorists. The cyber-terrorists of the future may be supplemented and even co-opted by geomagnetic storms to give a whole new meaning to the term "infrastructure collapse."

Cyrogenic: The specialized heat treatment of tool steels, including woodworking edge tools, by freezing to increase durability and hardness, by soaking at a temperature of -320 °F for up to 20 hours followed by double tempering. Cyrogenically treated steel is now being used by contemporary toolmakers such as Lie Nielsen, Hock Tools, Bridge City Tool, and Knight Toolworks.

Damascene: "a. to ornament (metal work esp. steel) with designs inlaid in the surface and filled in with gold and silver. b. to ornament (steel) with a watered pattern, as in Damascus blades" (Oxford English Dictionary 1971, 641). Stone (1999) defines damascening as the process of inlay where the design is chiseled onto the surface of the item and the precious metal (gold or silver) is hammered into the cavity and is disclosed on the surface of the sword as either a flush or relief pattern. Stone notes the process of false damascening in the English tradition, which an anonymous contemporary source has described as: "Steel that has been overlaid with designs of precious metal anchored in a cross hatched pattern that has been pricked in the surface of the steel (as opposed to inlayed)" (personal communication 2008). Smith (1960) defines damascene as the treatment of a pattern-welded iron and steel sword by a reagent for the purpose of disclosing the crystalline pattern of the microstructure of the sword, which results from the forge welding of the iron and steel and is dependent upon the unique combination of iron and steel constituents and the lattice structure of the material being forged.

Damascened steel: Steel, especially steel swords, ornamented by designs inlaid in the surface and filled in with precious metals *or* treated with a reagent to disclose microstructural patterns. In the case of Japanese swords, makers can often be identified by their characteristic patterns, which appear after treatment with the acid reagent and express their forge welding techniques. In the 18th and 19th centuries, Englishmen collected damascened gun barrels. The patterns disclosed by the use of the reagent illustrated the artful forging technique of the gunsmith or swordsmith. The art of damascening by inlay requires no reagent. Damascened steel is not the same as Damascus steel. See Smith (1960) for commentary on damascened versus Damascus steel.

Damascus steel: Specialized form of crucible steel made for producing pattern-welded wrought iron and steel sword blades; made in India (Wootz steel) and Asia (Asian crucible steel), usually in very small batches of 2-3 kilograms per firing and then transported as a trade item to Near East Asian communities, such as Damascus, for reforging into swords; possibly one of the earliest forms of steel. As Wootz steel, it was known in Europe after 1500 but was not widely utilized due to the difficulty of reaching a temperature sufficient to relieve its hard, brittle, crystal structure by reforging. See Wootz steel, Stourbridge clay, Moxon, Damascus blade, and cast steel.

Damascus sword: Term used to describe Persian, Indian, and Indo-Persian swords, daggers, and scimitars, examples of which survive from the 16th to the 19th century and are characterized by a diversity of patterns based on alternating layers of high and low carbon steel but dissimilar to the traditional pattern-welded Merovingian sword. Damascus blades (swords) were made out of Wootz steel. According to Smith (1960) "Attempts to duplicate Wootz and watered steels generally led to James Stodart's work in France early in the nineteenth century, most of which was based on the erroneous belief that the texture arose from forging. Many workers described the production of fancy surface patterns by welding, twisting, and forging together variously shaped pieces of iron and steel. The results were both decoratively and mechanically inferior to the Oriental product and, in fact, were merely a return to sword-making of a technique descendent from that of the Merovingian smiths which had meanwhile been extensively used in the Near East and India for making so-called 'Damascus' gun barrels, which had nothing but name in common with the sword." Recent research (Levin 2005, etc.) indicates that the Wootz steel used in Damascus swords has a high but heterogeneous carbon content (1.5 - 2.0% cc), may contain micro-constituents such as cobalt, vanadium, and chromium, and was probably forged at temperatures below the critical temperature, creating carbon nanotubes, which provide Damascus swords with their characteristic superplasticity. One probable reason why the production of Damascus swords cannot be replicated is the depletion of the ores of continental India, from which they were derived. See superplasticity and Wootz steel.

Damask: Etched pattern. When present on the surface of Merovingian and other pattern-welded swords, the damask expresses the crystal structure of the pattern-welded iron and steel in the sword. See damascened steel.

Damping: The capacity of a metal to absorb vibrations, reduce noise, and lower stress by converting mechanical energy into heat. Grey cast iron has an elevated damping capacity due to the presence of flake graphite in its microstructure.

Dapping block: A small rectangular steel block characterized by indented circles of various diameters and depths used by a silversmith to shape silver. Often 4" x 4" x 4", the dapping block is a miniature version of a swage block but without its rectangular slots.

See swage block.

Debt: The totality of mortgage debt, college loan debt, credit card debt, consumer debt, corporate debt, bond debt, pension fund debt, US government debt, banking debt, sovereign government debt, including foreign debt. Total global debt in 2012 exceeded 190 trillion dollars. In contrast, total world gross productivity was about 80 trillion dollars. The shadow banking network makes huge profits selling and trading debt as an asset, often in the form of collateralized debt obligations. Its shadow banking network debts as assets now exceed the total amount of both gross world productivity and gross world debt. The Age of Biocatastrophe is also the Age of Klepto-plutocracy based on the predatory but profitable trading in debt as an asset. Nearly invisible communities of tool wielders, artisans, artists, craftspersons, born-again ironmongers, and a minority of the elite technicians of our hyper-digital milieu have the challenge of formulating sustainable lifestyles in the context of the unfolding tragedy of a world commons characterized by debt as the primary asset of the klepto-plutocracy and its shadow banking network.

Decarburization: Removal of carbon from cast iron during fining or puddling or from steel during heat treatment by oxidation, the latter of which produces surface decarburization. Decarburization occurs either directly in an oxidizing atmosphere or indirectly with iron oxide as an intermediary, producing carbon monoxide or carbon dioxide gas. "Loss of carbon from the surface of heated steel. Decarburization prevents steel from being properly hardened" (Gordon 1996). The decarburization of cast iron to produce steel was the principal steelmaking strategy of the Renaissance and was not replaced by the alternative method of carburizing wrought iron to make blister steel until after 1690, and then primarily in England. See German steel, blister steel, and puddling furnace.

Decalescence point: The lowest temperature at which a change of structure occurs in steel (1341° F), at which time the body-centered cubic unit cell of gamma iron changes to the face-centered cubic unit cell characteristic of alpha iron accompanied by the absorption of heat. See austenizing temperature.

Deep hardening: Creating modern cutting tools, dies, and lathe centers that retain hardness at high temperatures and have a high resistance to wear and abrasion as a result of the addition of tungsten, chromium, vanadium, and cobalt (T type alloy steels). In M type alloy steels, molybdenum is substituted for tungsten and then combined with chromium or vanadium.

> The typical molybdenum high speed tool steel contains about 4 to 8.5 per cent of molybdenum, 4 per cent chromium, 1.5 to 6 per cent tungsten, and about 2 per cent vanadium. Since this material is very susceptible to oxidation, these tool steels must be heated in a fused salt bath or coated with borax. They must be pre-heated to approximately 1400 F and then heated quickly to 2200 F. They are quenched in air, oil, or fused salts, depending on the type of steel used. They also have a strong tendency to decarburize. These steels may be tempered at 1050 F. (Pollack 1977, 219)

Deforestation: The systematic removal of the mature forests of terrestrial ecosystems for the purpose of agricultural or industrial activities or for the construction of human habitations. Deforestation was the fate of large areas of North American forest after the arrival of colliers producing charcoal fuel for bloomsmiths, blast furnaces, foundries, and other ironmongers. Deforestation also occurred in Mediterranean, Near Eastern, and European forested areas in the early Iron Age, and is currently occurring in tropical forests in the twilight years of the pyrotechnic society and its systematic exploitation of the earth's natural resources. The recent global demand for biofuels is now increasing both the rate of deforestation and the annual atmospheric levels of CO_2 and other greenhouse gasses that cause global warming. See collier.

Deformation: Changes in form produced by external forces and/or temperature change acting on the non-rigid microstructures of metals, including ferrous metals. See creep, lattice structure, microstructure, and stress.

Deformation bands: New grain boundaries that result from the rotation of the lattice structure of metals during slip events, which may occur during cooling, tempering, annealing, forging, or cold hammering. See creep, lattice structure, microstructure, and slip.

Deformation history: Record of mechanical and heat treatment retained in a piece of metal, exhibited by shapes of grain and slag particles. For example, cold working by hammering leaves telltale Neumann bands in ferrite grains. See lattice structure, microstructure, and stress.

Deformation twinning: The fractural movement of the atomic planes of crystal lattice structures over each other, in contrast to the whole number increments of slip deformation, during the microstructural deformation processes caused by the thermal treatment of austenized steel (Pollack 1977, 73-75). It is a common mechanism of plastic deformation in pearlitic steels. See twinning.

Dendrites: Branching forms characteristic of crystal growth patterns in metallurgy. See crystal structure and crystallography of ferrous metal.

Deoxidation: The removal of oxygen by the addition of elements with a high oxygen affinity, e.g. carbonaceous materials, as in the smelting of iron oxide ore and its reduction to alpha iron or in crucible steel production to suppress formation of gas porosity during the solidification of the liquid steel.

Deoxidizing agent: Elements used to remove oxygen or oxides from metals being smelted, e.g. silicon or aluminum.

Depth of case: Depth of hardness penetration sought or achieved during the case hardening of steel tools. See case hardening and case carburization.

Diderot's *Encyclopedia* ([1751-75] 1959): Widely circulated multi-volume encyclopedia illustrating important French trades, such as carriage making. The Diderot *Encyclopedia* is particularly detailed in its depiction of the blast furnace, anchor and cannon forging, casting of statutes, mining and smelting of metals, and manufacture of ornamental iron work, which was a specialty of the French. The encyclopedia provides vivid pictorial evidence of the longevity and the continental European origins of many of the tools that reappear in New England workshops from the colonial period to the mid-19th century, including the leg vise, anvil, and other related tools that would have been used by toolmakers making their own hand tools. First published in the 18th century, the encyclopedia is currently available in multiple reprinted versions. Sixty four plates of hand tool illustrations from the *Encyclopedia* are reprinted in *Tools Teach: An Iconography of American Hand Tools* (Brack 2013). See leg vise and pattern books.

Die casting: Shaping of molten metal by the use of external pressure, as in drop-forging.

Die driving: Hammering, heating, and tempering of dies prior to final casting, drop-forging, or embossing of die-struck patterns or models. During hardening and tempering, all dies must be shielded from the oxidizing effects of the fire by charcoal or a sheet iron cover.

Die-sinking: The art of creating the patterns for drop-forged tools, which are hammered or hydraulically forged out of hot carbon steel, tempered, annealed, and sometimes subjected to additional mechanical and thermal treatments, including re-tempering to restore hardness. Drop-forged dies are created by taking a pair of forged crucible blanks, annealing them for softening, and shaping them by filing, chipping, and machining (boring, slotting, and milling). The machining of the exactly matched die patterns provides a surface that can be finished by polishing and further machining the edges. The matched dies are re-dropped (reheated for tempering), pickled to remove scale, and annealed again at low temperatures before being surface hardened for the fluid die-casting of drop-forged tools and other artifacts. During the die-sinking or shaping phase, dies are often inserted into cast iron die holders. The best die steel alloy (c. 1910), superior to carbon steel, was vanadium steel (Woodworth 1911). Occasionally, especially in the 19th century, cast iron dies were used for some forgings. See die driving, drop-forging, and drop press.

Die-sinking machine: Machine that replaced the hand work of the die-sinker and enhanced production uniformity of interchangeable dies.

Direct conversion method: Method that allowed the direct conversion of Swedish charcoaled bar and pig iron, in combination with quantities of scrap iron, into crucible steel without the use of the converting (cementation) furnace; invented by William Vickers in Sheffield in 1839. By 1880, most use of converting furnaces to produce blister steel had ended. See blister steel.

Direct process: Single stage production of wrought and malleable iron from iron ore in bowl and low shaft bloomery furnaces; also, an appropriate description for the single stage production of hot-forged natural steel edge tools by a blacksmith directly from the bloom, rather than from iron bar stock. Direct process iron is often high in slag. See bloomery, wrought iron, malleable iron, and blast furnace.

Displacement: Slip or creep of the crystal lattice structure of metals during their plastic deformation. See slip, creep, plastic deformation, and lattice structure.

Double-acting steam engine: A steam engine designed to allow steam to be admitted alternately to each side of the piston while the other is exhausting due to the design improvements of valve-controlled steam flow and the use of inlet and exhaust ports at either end of the cylinder. Designed and produced in and after 1783, James Watts' double-acting steam engine, soon called the reciprocating steam engine, represented a major improvement in the efficiency of his first steam engines, which now could be mounted at any angle and operated with increased speed and smoothness of reciprocation. The reciprocating steam engine became widely used in the English textile industry after 1785.

Double-process quenching and tempering: Modern, more complicated form of heat treatment associated with relatively easy to melt high carbon steels; particularly useful for the manufacture of super-hardened alloy steel composites.

Dozzle: Clay cone invented by R.F. Mushet in 1861 to contain a reserve of liquid steel to alleviate the pipe in crucible steel ingots. The dozzle was inserted into the top of the ingot near the end of the pour (Gordon 1996). See pipe, crucible steel, and cast steel.

Drag: Bottom half of a foundry mold. See cope.

Drawing furnace: Furnace used for tempering axes after hardening. Typically an ax is heated to the color "pigeon blue". Rapid cooling fixes the temper of the ax (Kauffman 1972). See temper.

Drawing temper: Heating of hardened steel to a specific temperature, followed by its quenching to obtain a lesser degree of hardness.

Draw plate: "A plate of hardened steel, furnished with a series of tapered holes of gradually diminishing diameters through which metals are drawn out into wires" (Audel 1942, 200).

Drop-forging: Use of top and bottom dies to hammer a heated bar into the shape of a die sunk pattern. Hydraulic forging performs the same function by pressing heated metal into die sunk patterns. The drop-forging of hand tools began with the adaptation by the Collins Axe Co. of some of the inventions of the English Industrial Revolutionaries for ax and other tool production in the 1830s. The perfection of the production of malleable cast iron and grey cast iron in the 1850s and the production of tempered alloy steels after 1870 provided the basis for the widespread use of drop-forging technology in the American factory system of the mass production of tools and equipment of every description. Modern CNC technologies continue to use drop-forging equipment. Drop-forging techniques have their origin in the Bronze Age and early Iron Age where individual axes and knives were forged in molds with hammers before being heat treated and/or steeled by ancient ironmongers. See die-sinking and die driving.

Drop press: Machine for punching, shaping or embossing metals; originally hand or foot operated utilizing a vertically guided weight; later, water or steam powered; fundamental machine used for drop-forging tools. The most modern form of the drop press is the hydraulic press. See drop-forging, die-sinking, and die driving.

Dross: Slag-like, metallic oxides that rise to the surface of the bath during the oxidation process, as in the reduction of iron ore in the blast furnace.

Dry puddling: Direct production of wrought iron by the conversion of white pig iron in a reverbatory furnace. Conversion of grey cast iron to puddled wrought iron required preliminary refining due to its free carbon and low silicon content. See puddling furnace and wrought iron.

Dry sand molding: Casting with a flask and a baked sand pattern, facilitating moisture removal and providing a hard clean surface for detailed casts. See cope and drag.

Dubbing: Shaping and smoothing of frames and timbers with an adz during the construction of a wooden ship.

Ductile cast iron: A cast iron characterized by a microstructure with graphite in ball-like forms in contrast to the flakes of ordinary grey cast iron. Using special processing and magnesium and/or cerium containing alloys results in increased strength and ductility due to the nodulizing of the graphite. Widely used by the automotive industry, it is a modern variant of malleable cast iron.

Ductile steel: "Sir Joseph Whitworth's last contribution to mechanical engineering was his discovery that guns made from ductile steel would wear by losing their shape whereas hard steel guns tended to explode when they were unsound. The problem was in casting ductile steel into ingots without creating air pockets, which made the metal unsound. Sir Joseph Whitworth's solution was an adaptation of Bessemer's principle of hydraulic pressure casting." (The Whitworth Society http://www.whitworthsociety.org/history.php?page=2).

Ductility: Capacity of a metal to permanently deform without fracture when cold worked.

Dunnage: Iron fragments, scrap, and faggots used as a shim, as well as ballast, in the bottom of a wooden ship to raise the cargo above the bilge water level to prevent damage.

Duplexing process: Refinement of cupola-produced cast iron in an accurately controlled electric furnace for the purpose of producing special purpose, high quality, structural cast iron.

Early steel: Typically, natural steel produced in ancient and medieval times directly from the bloom. The carbon iron microstructure of early steel always contains elongated or fragmented silicon slag inclusions. See bloomery, Catalan furnace, ferrosilicate, and natural steel.

Ecotoxin: Naturally occurring and anthropomorphic chemical toxins, such as methylmercury and persistent organic pollutants (POPs), which originate at specific point sources and are then subject to regional and hemispheric transport by the biogeochemical cycles of the chemosphere. Many ecotoxins, such as biologically significant chemical fallout, bioaccumulate in pathways to human consumption. Iron and steel production facilities make a modest contribution to global ecotoxin contaminant pulses in the form of polycyclic aromatic hydrocarbons (PAH) and mercuric sulfide emissions from coal/coke fuel. See biocatastrophe, biologically significant chemical fallout, and global contaminant pulse.

Edge carburization: Alternative process for the carburization of iron, used by both early and modern steelmakers from the early Iron Age to the modern era. Prior to the advent of the mass production of edge tools, an iron tool would be enclosed by clay or another protective covering and submerged in a charcoal fire to allow diffusion of carbon to steel the cutting edge of the tool. Modern forging strategies subject the bit of an ax to additional mechanical and thermal treatments, including grinding, tempering, and annealing to increase the hardness and ductility of the cutting edge. See case hardening and enclosure.

Edge tool forging: Creation of steel edge tools by the heating of iron-carbon matrixes (FeC, Fe_3C) into the austenite phase followed by rapid quenching, which creates martensite (iron with homogenous carbon distribution), which is then tempered to relieve brittleness and then shaped by hammering, rolling, or casting into appropriate forms. Properties of edge tools can vary widely, depending on carbon content, carbon distribution patterns, mechanical processing (splitting, folding, piling, hammering, and rolling), thermal treatments, (including duration of quenching, tempering, and annealing), alloy content, if any, and slag inclusions and distribution. The key to successful edge tool production is the rapid quench, which forms martensite instead of pearlite, the latter of which is formed by slow cooling. This is followed by the slow tempering of the steel tool over a longer time and at a much lower temperature (+/- 600° C) to alter the acicular cementite microstructure of martensite to more finely and homogeneously distributed cementite spherical formations, which soften it and increase its ductility. As an alloy, silicon also functions as a softening agent, enhancing the ductility of the steel being forged. Edge tools of a more inferior quality could also be made by forge welding unquenched, slowly cooled austenite, but a sophisticated knowledge of the necessity of the thermal treatment of forged steel to make knives, swords, and woodworking edge tools dates to the early Iron Age (Barraclough 1984a).

Edge tools: Hand tools used by woodworkers to cut and shape wood, especially carving tools, chisels, gouges, slicks, adzes, axes, and plane blades. In the broader context of the history of hand tools, edge tools also include knives, swords, and cutlery. See convivial tools.

Elastic deformation: Less than permanent deformation, which is remedied by the removal of the stress or load causing the initial deformation.

Elastic limit: Point at which a metal becomes permanently deformed under stress; also defined as yield strength.

Electric arc furnace: Furnace in which the heat needed to smelt metals is produced by an electric arc between carbon or graphite electrodes and the furnace charge; also called a low frequency induction furnace. One of the two most important 20th century steel-producing strategies, made possible by an electric power grid, the prime mover of which was biomass-fueled steam turbines or water-powered turbines. Its special advantage is that, in the production of high grade tool steels, the oxidation of alloys, such as chromium or nickel, are avoided. The great majority of the hundreds of varieties of useful high grade alloy tool steels produced in the 20th century were manufactured in this furnace. Its appearance was significant for edge toolmakers because the steel produced in the electric arc furnace gradually replaced the crucible steel production process after 1900. By 1940, the disappearance of both the puddled iron process and the crucible steel process seems to coincide with the diminishing quality of edge tools. The electric arc furnace can be built in any size and utilizes scrap iron and steel efficiently. The first electric arc furnace was constructed by Sir Charles William Siemens in 1878 but was used for melting only. The first smelting of iron ore occurred in 1898. In 1927, the first high frequency electric furnace was constructed. Modern electric arc furnaces produce almost all steel used for contemporary toolmaking. Induction, consumable-electrode melting, and electro slag furnaces are three modern subtypes of the electric arc furnace. The induction furnace, which now has many variants, is a primary source of tool steels. The consumable-electrode furnace and electro slag re-melting furnace are used for production of special alloy steels and ultra high strength missile and aircraft steel. By 1980, electric arc furnaces were producing about 28% of all US steel. The uniformity and the high quality of alloy steels produced in the electric furnace is illustrated in the wide variety of sophisticated products of an atomic age, which was and is characterized by a diminished capacity for mass production of high quality edge tools. See basic oxygen process and furnace types.

Electro fusion: Modern form of steel production, which uses an electric current to melt the charge of ore and flux, c. 1898. In the early years of the electric arc furnace, edge tools made by this process were sometimes marked "electro fusion."

Electron microscope: Type of microscope that uses electrons to illuminate and create an image of a specimen. The electron microscope has much higher magnification and resolving power than a light microscope, with magnifications up to about two million times compared to a light microscope's about two thousand, allowing it to see smaller objects and greater detail in these objects. Unlike a light microscope, which uses glass lenses to focus light, the electron microscope uses electrostatic and electromagnetic lenses to control the illumination and imaging of the specimen. The three types of electron microscope are: the transmission electron microscope (TEM), the reflection electron microscope (REM), and the scanning electron microscope (SEM). See REM, SEM, TEM, and HRTEM.

Elongation: Permanent extension of the microstructure of a metal by mechanical processes, such as hammering or rolling, especially effective at high temperatures. In alloy steels, elongation is due to grain boundary sliding of alloys in tension, which then provide qualities of plasticity and superplasticity. See deformation bands, superplasticity, and nanotubes.

Embossing: Ornamental designs executed by stamping or hammering on the reverse side of a metal to form a relief. See chasing.

Embossing sequence: The first step in the embossing sequence is the creation of a plaster or wax casting of the pattern to be embossed, e.g. a medallion. A cast iron pattern is made from the plaster or wax pattern and is driven into the die to form a die-sunk casting. This casting of the original pattern then recreates the original design. See die casting, cire perdu, and die scaling.

Embrittlement: In ferrous metallurgy, a common result of sudden cooling, as in the formation of martensite; usually due to the formation of acicular cementite microstructures at grain boundaries, in contrast to spheroidization of cementite, which characterizes microstructures with more ductility and plasticity.

Enclosure: Ancient practice of protecting forged iron tools being carburized from the oxidizing influence of combustion gasses by covering the iron with a layer of organic materials (pig fat, goat skins, clay, etc.). For the production of sheet steel from sheet iron, the layer of charcoal served as the enclosure; for the production of Brescian steel, the liquid cast iron provided enclosure. Various methods of case hardening necessitate strategies of closure, e.g. the production of blister steel is based on the enclosure of iron bar stock in the protective layers of the sandstone converting furnace. See case hardening, carburize, and cementation furnace.

Energizers: Carbonates, which hasten case hardening and steel cementation, including the carbonates of calcium, barium, and sodium in commercial carburizing compounds. In early cementation furnaces, naturally occurring energizers in charcoal dust, bone, skins, and hide inserted between iron bar stock facilitated carburization. See case hardening and carburize.

Engine: Compound machine that uses any power source (water, wind, biomass fuels, heat, electricity, etc.) to do work. See prime mover.

Eoliths: First tools used by humans; un-hafted, un-knapped, found stones.

Erzberg: Name of the ore-bearing mountain in the Styrian section of Austria that was an important source of manganese-laced siderite iron ores used from the early Iron Age to the early modern period. The manganese content of these ores facilitated production of natural steel in early shaft furnaces, which was then traded throughout central Europe as currency bars. The manganese content also facilitated the production of high quality German steel by assisting in the removal of sulfur during the decarburization of cast iron made with Erzberg ore. See natural steel, Styrian steel, iron road, and currency bars.

Etching: Application of a liquid reagent to the polished surfaces of metals for the purpose of revealing their microstructures. This method was widely used after 1885, instead of visual analysis of a fracture, for the evaluation of the structural properties of ferrous metals. See microstructure and lattice structure.

Eutectic: Minimum melting temperature of an alloy, i.e. the temperature at which the crystalline becomes molten and the molten becomes crystalline but always involving a mixture of two or more components. In ferrous metallurgy, the most important eutectic point is the melting temperature of austenite at 723° C (1341 °F). Steel and cast iron are mixtures of austenite and cementite, which, if cooled below this temperature, become mixtures of ferrite and cementite. If, however, austenite is suddenly cooled in water from a red heat, martensite, (steel with uniform carbon content,) is produced. Hard and brittle, martensite is softened and made more ductile by tempering, which causes the dispersion of fine particles of cementite within the austenite. Another definition of eutectic with respect to cast iron is: "a thermodynamic equilibrium of two different solid phases with a liquid solution, e.g. an iron carbon mix with 4.3% carbon content (the eutectic composition) at 1150° C (the eutectic temperature) where liquid iron carbon is in equilibrium with austenite and cementite" (Wayman 2000). See ice cream, austenite, carbon content, and cementite.

Eutectic composition: In ferrous metallurgy, "…a thermodynamic equilibrium of two different solid phases with a liquid solution" (Wayman 2000).

Eutectic point: Minimum melting temperature of an alloy.

Eutectoid: The thermodynamic point equivalent to a eutectic point but with all three phases solid, i.e. one solid solution and two solid phases, as in the iron/carbon system at 723° C (the eutectoid temperature) and 0.8% carbon (the eutectoid composition), where austenite is in equilibrium with ferrite and cementite (Wayman 2000). Tool steels with a carbon content of 0.6% or greater are examples of solid compositions, which have undergone a eutectoid transformation, also called the martensitic transformation. In the eutectoid transformation, the slow cooling of austenite produces ferrite and cementite (iron carbide), which often exhibit lamellar structures, including bainite and pearlite. See lamellar and eutectic.

Eutectoid composition: "…the composition which transforms wholly to austenite at the lowest possible heating temperature" (Bain 1939, 31).

Eutectoid steel: Steel that contains 0.83% carbon is known as eutectoid steel, i.e. austenite (Shrager 1949). Austenite with a carbon content of >0.83% is hypereutectoid; if containing < 0.83% cc, it is hypoeutectoid steel. See hypereutectoid and hypoeutectoid steel.

Eutectoid transformation temperature: The eutectoid transformation temperature varies with carbon content. The lowest eutectoid transformation temperature or eutectic point is 723° C (1340° F), at which mixtures of ferrite and cementite with a carbon content of 0.83% undergo phase transformation to austenite. Mixtures with both higher and lower carbon content undergo phase transformation to pure austenite at higher temperatures.

Explosive compaction: In powdered metallurgy, the high pressure compaction of metallic powders, composites, and ceramic and super-hard materials, including nanocrystalline aggregates, by the use of shock waves. See powdered metallurgy.

Eyepin: Insert used as the core around which two pieces of mild steel are wrapped in ax-making. A high carbon steel insert is then welded into the fold to form a steel ax. In 19th century ax-making factories, the hands of the grinder held the eyepin during finishing of the ax prior to the introduction of automatic grinding machines. "No ax grinder ever died from old age" (Kauffman 1972). See ax-making techniques.

Face-centered cubic structure (FCC): Location of atoms in a crystal structure at the corners and in the middle of a cubic cell, as in austenized steel. At a temperature of 723° C, the body lattice structure of alpha iron (BCC) begins deforming into the face-centered cubic (FCC) structure of gamma iron (austenite). The change is completed at 910° C/1670° F, when the lattice structure of the iron becomes a face-centered lattice structure. Upon further heating, the FCC structure of gamma iron reverts to the BCC lattice structure (delta iron) at a temperature of 2552° F/1400° C. See lattice structure, space lattice, and cubic units.

Facings: Graphite, sea coal, or other material applied to or mixed with the sand mold into which a metal is to be poured to provide a smooth surface for the casting. For steel castings, silica flour mixed with molasses water was used c. 1910.

Faggoting: Bundling of iron rods or rolling of puddled iron and steel plates into bar stock prior to further heat treatment and/or forge welding. See piling and billet.

Farga: Spanish word for the technology of the reduction of iron ore in direct process Catalan furnaces. This technology was known and used in Germany at least as early as the 16th century and was transported to Pennsylvania by German immigrants during the 18th century. Bloomsmiths in colonial New England may have been influenced by German immigrants, who brought their knowledge and interpretation of direct process Catalan bloomeries to England in the 16th and 17th centuries, after which it was transferred to New England in the great migration.

Fargue: Spanish name for direct process Catalan bowl furnace.

Farrier: Blacksmith who specializes in the shoeing of oxen and horses.

Fast cooling rate: Rapid cooling of austenite that prevents excess carbon from precipitating in the form of cementite. Iron atoms are rearranged to form an acicular crystal structure that traps carbon in a diffusion-less solid state transformation (martensite), in which no movement of microstructural dislocations is possible, thus producing a hard and brittle microstructure.

Fatigue: Progressive loss of resistance of a metal to fracture, usually occurring at ordinary temperatures, due to the application of force on a flaw that forms a crack that spreads under repeated stress. See plastic deformation and lattice structure.

Fayalite: Form of iron silicate formed as fluid slag in the smelting process by the mixture of silicate and iron oxide. Direct process bloomeries wasted significant quantities of iron ore as fayalite before the role of lime as a flux, which reduced the loss of iron oxide, was known. See acid process, basic steelmaking process, flux, and slag.

Feedback mechanism: Unintended adverse consequences of cascading Industrial Revolutions, e.g. the impact of the combustion of fossil fuels by pyrotechnic industrial society produces CO^2 as a greenhouse gas, which warms the atmosphere, melting permafrost, which releases large quantities of methane, another greenhouse gas, all of which cause sea level rise by melting glacier ice. Can born-again ironmongers using nanotechnologies invent some substance that could be used to build 40 foot high seawalls around vulnerable cities to reduce the impact of cataclysmic climate change (e.g. Hurricane Sandy)?

Feeders and risers: Fittings utilized during the solidification of cast metals in foundry work to offset shrinkage by allowing the addition of molten metal during the casting process.

Female die: Lower die used in press working of jewelry, silverware, and toolmaking; also known as embossing die. See die-sinking.

Ferdoux: French term for wrought iron produced by decarburizing cast iron. See acier fondu, acier forge, ferfort, and wrought iron.

Ferfort: French name for malleable iron produced by decarburizing cast steel. Ferfort has the same carbon content as modern "low carbon steel," thus differentiating it from wrought iron, which the French call *ferdoux*. Ferfort was frequently used in the manufacture of horticultural tools (Barraclough 1984a). See ferdoux, German steel, low carbon steel, malleable iron, and wrought iron.

Ferrite: "Low-temperature form of pure iron" (Gordon 1996). Ferrite is a solid solution of carbon in iron up to 0.03% (alpha iron) stable below 723° C, as in wrought iron. "It is the softest component in steel and is very ductile" (Pollack 1977, 124-5). Ferrite contains trace amounts of iron silicide (FeSi) and iron phosphide (Fe_3P). It is strongly magnetic, soft, and ductile and has a tensile strength of between 40 and 50 psi and a Brinell hardness of +/- 80. See austenite, cementite, and wrought iron.

Ferrite-carbide aggregate: The mixture of pure iron (ferrite) and iron carbide (cementite) that is characteristic of many forms of carbon steel. The iron carbide content is dispersed in a ferrite matrix. The proportion of cementite in steel or cast iron depends on carbon content. The wide variety of microstructural forms of ferrite-carbide aggregates in pure carbon steel is a function of heating and cooling rates and variation in thermo-mechanical treatments: hammering, tempering, annealing, etc. See austenite, bainite, carbide, cementite, ferrite, lattice structure, and pearlite.

Ferrite-cementite phase: Eutectoidal mixture of alternating lamellae of ferrite and cementite, commonly called pearlite.

Ferrite-cementite-martensite complex: Specific carbide microstructure common to many steels and also associated with superplasticity in steels. X-ray diffraction analysis of the lateral face of a Wootz steel Damascus sword blade indicates a microstructure of α-Fe (ferrite) (73.8% wt), Fe_3C (cementite), and $Fe_{1X}C_X$ (martensite), without the presence of either austenite or pearlite. The blade face also contained 2.4% wt carbon content in the form of graphite (Levin 2005).

Ferrite-Pearlite (FP) Structural Steel: The most industrially significant form of modern structural steel characterized by small ferrite and austenite grain size in the 5-6μ (micron) range. The appropriate heat treatment of ferrite-pearlite steels creates a matrix with virtually no undissolved carbides. The refinement of the ferrite and austenite grain size creates a resistance to impulsive loading stresses (Paxton 1967, 3).

Ferritic malleable cast iron: Malleableized cast iron with a ferrite matrix imbedded with particles of temper carbon.

Ferromanganese: Alloy of iron and manganese containing 80% manganese, 12% iron, 6.5% carbon, and 1.5% silicon, which is added to the steel during the smelting process to increase its carbon and manganese content. Ferromanganese played a key role in upgrading the strength and durability of Bessemer's bulk process low carbon steel to make it suitable for uses such as steel rails. Ferromanganese also serves as a deoxidizer. See Bessemer process and manganese.

Ferrosilicate: Siliceous (silicon bearing) flux formed from unreduced ferrous oxide and fusible silicate slag that occurs in bloomery furnaces during smelting. Ferrosilicate encourages oxidation of carbon at the end of the heat in early furnaces and thus assisted in the production of wrought iron, preventing carbon uptake that would create unwanted liquid cast iron (Schubert 1957, Wertime 1962, 45).

Ferrous bainite: "All nonlamellar aggregates of ferrite in carbide (irrespective of ferrite morphology) as well as divorced pearlite" (Sinha 2003, 9.3).

Ferrous materials: Iron and iron alloy combinations within the iron carbon system.

Ferrous metals: Any metal containing iron. Most ferrous metals are iron-carbon alloys. Many other metals may be deliberately added as alloys to iron-carbon formulations. Numerous other chemicals such as phosphorus, sulfur, and silicon are inevitably present as micro-contaminants. See alloys, silicon, manganese, chromium, and rebar.

Ferrous metallurgy: Art and science of smelting iron from iron-bearing ores and the consequent use of hot iron by multi-tasking blacksmiths to forge iron and steel tools, weapons, and other implements.

Ferrum Noricum: Iron products produced in the ancient Roman province of Noricum, now southern Austria.

Fettling: Millscale (iron oxide) used to line a puddling furnace.

File-making: To create a file. Before the era of machine-made files, the process of making a file involved the cutting and cropping of "strings" of steel into "moods." File steel "moods" were forged into triangular or half-round blanks and then softened before surface scale was removed by filing. The cutting of the file edge with hammer and chisel was followed by hardening by quenching, cleaning, and oiling. File-making is thus anomalous, in that softening preceded hardening (Dane 1973).

Filing: Traditional method of shaping metal prior to the introduction of grinding machines (c. 1875). For example, the hand file was used to make parts that would fit a particular firearm before the factory system signaled the beginning of the mass production of interchangeable tool and machine components. The telltale marks of hand filing can be seen on many hand forged tools, which received their final shaping from hand filing by individual toolmakers.

Fin: Surplus iron or steel at the parting line of a two piece die from any forging, which is later trimmed off.

Fine grain: Uniform pattern of grain distribution in the lattice structure of iron-carbon alloys resulting from simultaneous crystallization of the lattice structure nuclei in contrast to random crystallization, which produces more deformed and irregular crystal structures.

Finery: Facility used for the decarburizing of pig iron to produce wrought iron using charcoal fuel; a hearth used to further refine bloomery-derived wrought or malleable iron. Products of the finery were often sent to a chafery for further thermal and mechanical treatment, including shaping into bar stock. Finery-produced bar stock from direct process bloomeries or the integrated iron works of a blast furnace complex was an important cargo in the coasting trade and the main source of iron for blacksmiths in the manufacture of hand tools. The finery was common before 1830 but was gradually replaced by the puddling furnace and specialized chafery pit-type furnaces for hot working or rolling malleable into semi-finished products such as slabs, billets, and blooms, the latter not to be confused with the blooms of the direct process of iron and steel production before the era of bulk steel production. See bloomery, malleable iron, puddling furnace, rolling mills, and wrought iron.

Fining: Process of purification by removal of contaminants, as in a fining furnace.

Fining hearth: **1**) Walloon and other open-hearth style furnaces used to decarburize cast iron into wrought and malleable iron at an integrated iron works. An increase in the fuel to ore ratio was sometimes used to produce malleable iron or raw steel instead of wrought iron, i.e. the decarburization of the cast iron was halted before all the carbon was eliminated from the pig iron being refined. In many cases, the iron being produced was malleable iron 0.08 – 0.2% cc or a malleable iron with a higher carbon content equivalent to that in modern (>1870) low carbon steel (0.2 – 0.5% cc). However, finery iron differed from modern low carbon steel in that it has a variable siliceous slag content, hence the term malleable iron. **2**) A secondary hearth for refining the wrought or malleable iron loup produced by the bloomsmith. Water powered North American colonial era smelting furnaces often had multiple hearths for further hammering and thermal treatment of iron bar stock. See cast steel 2), malleable iron, and Walloon process.

Firearm: Mechanical device that ejects a missile by the explosion of gun powder. The largest firearms were cannons, initially made from bronze, and, later, from cast iron. Smaller firearms, such as the Arquebus and the flintlock, were usually made of soft, folded, and welded sheet wrought iron. In the post-Reaganomics era, high powered assault rifles have become an icon of freedom as the influence of crypto-fascism has accelerated.

Flame hardening: Localized heat treatment of a steel surface to be hardened by the heating of the surface above the critical temperature (1300° F) to 1550° F.

Flange iron: Soft, ductile wrought iron used in applications subject to heavy stress.

Flash: Waste metal that remains on drop-forged tools after steam hammering.

Flask: Foundry molding box, consisting of the cope (top) and drag (bottom), often with a core as part of the pattern, into which sand is rammed and packed.

Flat world: A term coined by Thomas Friedman and used as a title in *The World is Flat: A Brief History of the Twenty-first Century*. Flat world refers to the capturing of the attention of a large percentage of our population, including students, workers, entrepreneurs, the educated elite, etc. by two dimensional Information Age technology, including computers, iPhones, iPads, and a growing repertoire of other electronic devices. Unfortunately, Friedman failed to reference, at least until later publications, the ongoing escalation of the three dimensional round world crises that characterize the evolving age of biocatastrophe: the world under crisis, catastrophic climate change, chemical fallout and its human health impact, growing debt and income disparities, declining public resources for infrastructure, education, and innovative research and development, political polarization and paralysis, and the social unrest these engender.

Flatting mill: Small hand-cranked set of iron rollers with an inset cut into the rollers to produce silver molding patterns, used by silversmiths on teapots and other wares in lieu of hand-stamped chased molding. Flatting mills appeared +/-1720 and were a miniature predecessor of later die-forging machinery of the factory system of tool manufacturing >1830. The moldings produced by the flatting mill are also called die-rolled moldings.

Flintlock: Dominant form of small firearms from the reign of William III (1650) until 1840 (Greener 1910). Developed in the 1620s and of Spanish or Dutch origin, the flintlock was much easier to fire than the matchlock. See matchlock, Arquebus, wheellock, gun barrel iron, and gunsmithing.

Flow-strain rate response: Measurement of flow stress (MPa) over strain rate (s^{-1}); "A strain rate sensitivity … exponent of 0.5 is often considered the optimum one for achieving superplasticity in fine-grained, metallic alloys" (Sherby 1995, 14).

Flow termination: Termination of plastic deformation in the crystal structure of metals that result from either brittle cleavage or ductile fracture.

Fluid carburization: Modern process for the hardening of steel by its heating above the critical temperature in a retort furnace in the presence of CO_2, ethane, methane, propane, natural gas, or liquid cyanide, any of which facilitate carbon penetration into the steel.

Fluidity of microstructures: Gestalt of diffusing, sliding, slipping, encompassing, encroaching, straining, elongating, shearing, and/or proliferating microstructures characteristic of the phase transformation kinetics of ferrous metals.

Fluidized bed process: Modern process for producing sponge iron that uses the heat from the partial combustion of preheated air and natural gas for the reducing reactions. "The reducing gas is composed of about 21 per cent carbon monoxide, 38 per cent nitrogen, and 41% hydrogen. The reduced iron is pressed into briquettes and stored or shipped" (Pollack 1977, 7). See sponge iron.

Flux: Additive in the smelting process, e.g. lime, which has the purpose of taking up and absorbing the melted impurities in the iron ore, creating slag. The slag is then drained off separately from the molten ore produced by direct process smelting or indirect process blast furnaces, usually through an opening above the container of the molten iron. Clamshells were the flux of choice in the bog iron forges and blast furnaces of southeastern Massachusetts. In modern steelmaking and ironworking, fluxes are used for bronzing to remove or prevent the formation of oxides and promote the free flow of filler material. They may have the form of pastes, powders, gasses, or other coatings and include borates, fluorides, chlorides, acids, and wetting agents. See brazing.

For-profit mental illness: Thorazine, Iproniazid, Prozac, Effexor, Paxil, Serzone, Celexa, Zoloft, Risperdal, Seroquel, Zyprexa: pharmaceuticals made by pharmaceutical companies are the prime movers of an epidemic in mental illness (Angell 2011). The convivial tool wielding lifestyles of round world born-again ironmongers provide an alternative to the flat world enhanced necessity of addictive but profitable pharmaceutical drugs.

Forest of Dean: Area just north of Bristol and the River Severn on the west coast of England, where low phosphorus brown hematite iron ore of nearly equal quality to Swedish ores was smelted in the 16th and early 17th century before being depleted. Along with iron smelted from ores located to the east in the Weald of Sussex, iron produced in the Forest of Dean may have played an important role in the manufacture of tools for the early colonial settlements in North America before the manufacture of blister steel became widespread (1700). The first blast furnaces in England were larger than those in the Weald and may have been located in the Forest of Dean (Cleere 1985).

Forge: (noun) See furnace/forge.

Forge: (verb) To shape steel and iron into various forms by hot working, including hammering, rolling, slitting, and bending. Until the early 19th century, forge welding by hand was the primary means of shaping iron and steel tools. Henry Cort's redesign of the rolling mill (1784) represented a radical improvement in the efficiency of the forging of hot iron. Beginning in the mid-19th century, most shaping or hot working of steel and iron was done by machinery that forced hot metal into dies and rolling machines. In the case of the manufacturing of hand tools, the patterns cut into the dies formed the tools being shaped. See drop-forging, die-sinking, and steeling.

Forge train: "The set of heavy rolls between which the shingle bloom of puddled iron is rolled into puddled bar (muck bar) The rolls are grooved in a diminishing series so as to reduce the bloom to a manageable bar, and are made reversing if two high, or non-reversing if three high" (Audel 1942, 246). See rolling mill.

Forge welding: Bonding of pieces of steel or iron by hammering at high temperatures, often in the presence of flux. See steeling.

Forging mill: Facility for producing finished steel by hammering and rolling. These mills were particularly efficient when run by a water-powered hammer, which had even strokes and more controllability than early steam-powered hammers, i.e. before the invention of reciprocating steam engines with double acting pistons and smooth even strokes. See trip hammer.

Foundry: Workshop or factory for casting metals, including grey cast iron, malleable cast iron, and nonferrous metals into cannons, hollowware, tools, hardware, and other equipment. Sometimes the foundry was attached directly to a blast furnace, and, thus, no re-melting of the metals was required. Foundries were often associated with cupola furnaces.

Fracture: Characteristic crystalline pattern (grain) exhibited after accidentally or deliberately induced cleavage altered or exposed the microstructure of the steel being thermo-mechanically processed. A fracture by brittle cleavage produces a flat, smooth surface. A shearing fracture produces a "rougher surface, usually slightly inclined to the slip plane….When this type of fracture brings ductile elongation to an end, it is found that the fracture occurs at a definite value of the resolved shear stress along the slip plane" (Barrett 1943, 321). See Sheffield classification of blister steel, grain, cleavage, crystal structure, and sap.

Fracture toughness: Ductility in iron and steel, characterized by the capacity to flex without breaking. See fracture, ductile iron, and ductility.

Free oxygen: Oxygen that in a furnace causes iron to burn; one of the hazards for a foundry master in situations where slag is not protecting the molten iron from oxidation.

Freezing point of iron: The freezing (or melting) point of iron is dependent on its carbon content (cc). Pure iron freezes and melts at 1535° C. As the carbon content of iron, steel, and cast iron increases, the melting and freezing point declines. Cast iron with a 4.3% cc freezes and melts at a temperature of 1130° C, the eutectic temperature of this alloy composition. Below the freezing/melting temperature, the iron or steel changes from a liquid (the liquidus) to a mushy combination of liquid and solid until the temperature falls sufficiently for the alloy to pass the solidus, or temperature at which it becomes completely solid. See ice cream and solidus.

French iron and steel: France utilized the traditional Continental technique of making "German" steel by decarburizing cast iron, sorting the results into three categories, ferdoux (wrought iron), ferfort (malleable iron), and acier fondu (quenchable steel that could be martinized, the highest quality of which is cast steel) (Barraclough 1984a). See acier forge and German steel.

Friction welding: Modern welding technique involving the joining, by spinning and jamming, of two pieces of metal in tubular applications. The high temperatures needed for welding result from the rapid spinning (30,000 – 40,000 rpm) of the metal being welded.

Fuel to ore ratio: Key consideration of early direct process smelting strategies. Differences in the fuel to ore ratio could result in the production of a low carbon wrought iron (< 0.08% cc), malleable iron with a carbon content (0.08% - 0.2% cc), malleable iron with a carbon content equivalent to modern (1870) "low carbon steel" (0.2 - 0.5% cc), or raw steel (> 0.5% cc). In the heterogeneous bloom of iron being smelted, increasing the fuel to ore ratio tended to increase the probability of raw steel production. See natural steel, carbon content of ferrous metals, and fining hearth.

Full annealing: Heating of iron alloys above their critical temperature for an appropriate time period, followed by slow cooling, either in a furnace or in a thermally insulated environment for the purpose of altering the crystal structure and, thus, the properties of the metal being annealed. See annealing.

Full welding heat: Temperature necessary to weld two pieces of iron or one piece of steel and a piece of iron together (2400° F / 1365° C). The iron will have a white color and will emit visible sparks due to the oxidation of the iron.

Fullerene: "Any molecule composed entirely of carbon, in the form of a hollow sphere, ellipsoid or tube… Fullerenes are similar in structure to graphite, which is composed of stacked graphene sheets of linked hexagonal rings; but they may also contain pentagonal (or sometimes heptagonal) rings… The discovery of fullerenes greatly expanded the number of known carbon allotropes, which until recently were limited to graphite, diamond, and amorphous carbon such as soot and charcoal" (Wikipedia 2013).

Fullering: Shaping of wrought or malleable iron by the use of a creaser (the fuller), which may be placed in a hardy hole to shape the iron or may be struck by a hammer to spread the iron by the formation of a series of indentations.

Furnace/Forge: Furnaces, including blast furnaces, are used for smelting ore or the melting of pig iron (cupola furnace); forges are used to reprocess the product of the bloomery or blast furnace by the mechanical and thermal treatment of metal after smelting. In some cases, especially before 1900, small forges also served as a furnace for reducing ore prior to the production of tools and ironware by the blacksmith or shipsmith. In town histories and historical writings, "forge" is a term often mistakenly used to refer to the furnace of a bloomsmith. The loups or blooms produced by the bloomsmith in his furnace were often shaped into bar stock in chafery or finery forges and transported to the forge of a blacksmith or shipsmith for further processing. See blast furnace, bloomery, chafery, ferrous metallurgy, finery, and fining.

Furnace bottom: Slag that formed at the base of a non-slag tapping bowl furnace.

Furnace types (before 1870): Three basic types of furnaces characterize iron and steel production from the early Iron Age to the beginning of bulk process steel production (1870), the crucible, bowl, and shaft furnace (Tylecote 1987, Wertime 1962). Later Stuköfen and blast furnaces are larger forms of the shaft furnace. Cementation, reverbatory, and cupola furnaces are further adaptations of this form.

> **Crucible-shaped furnace**: Among the earliest furnace designs was a simple crucible, especially common in northern Asia and used for both cast iron and Wootz steel production. Multiple crucibles with as little as 1 kg capacity were used for Wootz steel production as early as 300 BC. Cast iron production in larger crucibles in China can be dated at least as early as 800 BC. The crucible-shaped furnace evolved into the cupola furnace commonly used in the 19th century to re-melt blast-furnace-derived cast iron.

> **Bowl furnace**: Earliest form of furnace. The charge of ore was often located in back of the fuel and reduction occurred via a current of carbon monoxide formed by the burning charcoal, assisted by a directed flow of air from a tuyère. The Catalan furnace of northeastern Spain is the most well known form of the bowl furnace. Most earlier bowl furnaces were slag pit types. The rectangular open-hearth breakdown furnace used to smelt raw steel for Japanese swordsmiths is one of many variations of the bowl furnace form.

> **Shaft furnace**: Most common form of furnace throughout the Iron Age. Ore and fuel are mixed together in a four sided shaft. The resultant bloom of wrought or malleable iron is extracted from a hole at the bottom of the furnace. Combustion is also aided by the use of the tuyère, powered by varying types of blowing devices.

Shaft furnace types are further subdivided as being slag pit types and slag-tapped types.

>> **Slag pit furnace**: Furnace in which the slag accumulates *in situ* at the bottom of the furnace until the furnace is moved to an adjacent slag-free location.

>> **Slag-tapped furnace**: The slag-tapped furnace was more practical and efficient than the slag pit furnace, where the slag from the smelting process was withdrawn through a tap hole to an adjacent cavity or hollow in the ground.

> **Stuköfen furnace**: Low shaft (1 to 2 meters in height) slag-tapped furnaces characterized most Roman era ironworks. Low shaft furnaces gradually grew in height, capacity, and efficiency to become the high shaft Stuköfen of the German Renaissance, the immediate predecessor of the blast furnace.

> **Blast furnace**: A form of high shaft furnace designed specifically to operate at high temperatures, thus carburizing iron ore to produce liquid cast iron. Blast furnaces gradually increased in size from the high shaft Stuckofen furnace to the modern

furnaces with a capacity for converting up to 50 tons of cast iron to low carbon steel.

Two variations of the shaft furnace appeared during and after the 17th century, which were characterized by furnaces designed to prevent fuel-ore contact. Along with the modern blast furnace, these two furnace types dominated blister steel and puddled, wrought, and malleable iron production until the appearance of modern bulk steel furnaces. Both facilitated greatly increased quality control and quantity of steel and iron production.

Cementation furnace: Sandstone furnaces, in which iron bar stock was enclosed to be carburized into blister steel, also called "steel furnace," especially in the United States in the colonial period. First noted in Nuremburg in 1601, such furnaces protected the steel being smelted from the oxidizing influence of the burning fuel. The cementation furnace should not be confused with the many modern forms of steel furnaces developed after 1860.

Reverbatory furnace: A furnace form with a fire pit located underneath the hearth that facilitated the decarburization of large quantities of pig iron by heat reflected from the metal roof of the furnace, preventing fuel to ore contact. Primitive forms of the reverbatory (refractory) furnace, also known as the puddling furnace, were improved by Henry Cort into the modern form of the reverbatory furnace in 1785, allowing production of large quantities of high quality wrought iron.

See blast furnace, blowing devices, bowl furnace, shaft furnace, and tuyère. For modern furnace forms, see basic oxygen process, Bessemer process, electric arc furnace, open-hearth furnace, and Siemens-Martins.

Futtock: Four or five separate pieces of timber comprising the ribs of a ship, usually held together by iron bolts made by a shipsmith.

Galvanize: To coat the surface of iron and steel with tin or zinc, usually to prevent corrosion.

Ganister: Furnace linings made of refractory rocks, such as sandstone. See acid process.

Gangue: Contaminants in iron ore removed as slag, e.g. quartz and oxides of aluminum and silicon, often found in clay-like materials associated with the iron ore. The most common component of gangue is silicon. See ferrosilicate, slag, flux, and reduction.

Generosity of the 1%: The majority of the one percent in America, several million wage earners, doctors, actors, musicians, and successful entrepreneurs, are not members of the klepto-elite but rather fund charities and nonprofits that are a viable alternative to the stock and bond market casinos and tax loopholes of the American klepto-plutocracy.

Geoengineer: A person who measures and models the earth for applications using civil engineering technologies to combat, for example, the impact of climate change by producing temperature lowering atmospheric smog. Will innovative geoengineering concoct some creative solution to the looming disaster of increasing greenhouse gas emissions, cataclysmic climate change, and rapidly rising sea levels?

Geomagnetic induced current (GIC): Disruptions in the earth's magnetosphere and ionosphere caused by coronal mass ejection (CME) events, which cause voltage surges in power lines, transformers, pipelines, satellites, and other information technology networks.

Geomagnetic storm: A temporary disturbance of the Earth's magnetosphere caused by a solar wind shock wave and/or cloud of magnetic field which interacts with the Earth's magnetic field. The increase in the solar wind pressure initially compresses the magnetosphere and the solar wind's magnetic field interacts with the Earth's magnetic field and transfers increased energy into the magnetosphere. Both interactions cause an increase in movement of plasma through the magnetosphere (driven by increased electric fields inside the magnetosphere) and an increase in electric current in the magnetosphere and ionosphere. During the main phase of a geomagnetic storm, electric current in the magnetosphere creates a magnetic force which pushes out the boundary between the magnetosphere and the solar wind. The disturbance in the interplanetary medium which drives the geomagnetic storm may be due to a solar coronal mass ejection (CME) or a high speed stream (co-rotating interaction region or CIR) of the solar wind originating from a region of weak magnetic field on the Sun's surface. The frequency of geomagnetic storms increases and decreases with the sunspot cycle. CME driven storms are more common during the maximum of the solar cycle and CIR driven storms are more common during the minimum of the solar cycle (Wikipedia 2013).

German silver: Alloy of nickel and copper and sometimes zinc; also called nickel silver.

German steel: Steel made by the predominant steelmaking strategy in the Renaissance, both in England and on the European continent. German steel was made by decarburizing blast furnace pig iron in a separate finery type furnace (1400-1650). The resulting steel/iron mix was further refined into bar stock, bundled, reheated, quenched, broken up into fragments, reheated, annealed for +/- 45 minutes, and re-hammered as billets of steel (German steel). German steel was also produced in England after the establishment of the first blast furnaces in the mid-15th century until the cementation furnace became widely used after the mid-17th century. The roots of German steel production lie in the manganese rich ore of Styria and Carinthia (Austria), which facilitated the first iron production at Halstadt. Used by the Romans to produce steel weapons (Noricum), the possibly accidental production of natural steel in early bloomeries became the deliberate production of German steel in Stucköfen (shaft) furnaces, which could produce pig iron, steel, malleable iron, or wrought iron, depending on the fuel to ore ratio and smelting temperatures. Production of steel from decarburized cast iron in separate fineries was an outgrowth of the appearance of blast furnaces producing large quantities of pig iron in continental Europe after 1315. "Steel made from the seventeenth century onward in Styria and Carinthia by fining manganese rich pig iron. It contained alternating bands of high- and low-carbon content and was considered a superior material for cutlery" (Gordon 1996). Of great importance during the florescence of Nuremburg and Augsburg watch-making and armor manufacturing, +/- 1500 AD, numerous fine examples of hand tools made between 1500 and 1750 from German steel are on display at the National Museum in Nuremburg, Germany. Having the appearance of cast steel (Manganese in smelted iron ore results in a sheen, similar to that of cast steel.) and probably containing some martensite, these low carbon steel tools (+/- 0.5% carbon) are labeled as "eisen" (iron) by the National Museum. Their manganese content is the probable cause of their visual similarity to the crucible steel tools produced in Sheffield after 1750. In the broader context of tool production, between 1400 and 1800, German steel had a wide range of carbon content depending on its intended use. Many edge tools made after the appearance of the blast furnace, such as trade (e.g. Biscayne ax) and felling axes are composed of one piece of steel with no steel-iron interface, as in later "steeled" axes. See Biscayne ax, spiegeleisen, spathic ore, natural steel, and Styrian steel.

Gladius: Roman short sword made of wrought iron and low carbon steel, occasionally pattern-welded. The gladius was often cold work-hardened (not heat-treated) or surface carburized at its steeled edge. See spathas.

Glide plane: Points in the lattice structure that combine rotation on the screw axis with movement along a translation parallel, e.g. axial or diagonal length, of the lattice structure. Plastic deformation of the lattice structure of iron, for example, occurs along the crystallographic glide plane, also called the slip plane (Barrett 1943, 288).

Global contaminant pulse: The global transport of ecotoxins derived from industrial activities, transportation system and agricultural petrochemicals, and the electronic and pharmaceutical products of global consumer society. Global contaminant pulse transport occurs in the context of a biosphere composed of interconnected ecosystems where ecotoxins are detected as contaminant signals in abiotic and biotic media. These ecotoxins tend to biomagnify in pathways to human consumption and in human blood, tissues, and breast milk. See biocatastrophe, cascading Industrial Revolutions, ecotoxin, and tragedy of the world commons.

Goethite: "Ore mineral composed of hydrated iron oxide" (Gordon 1996).

Googolplex of microstructures: The endless variations in the microstructures of ferrous metals. If one "googol" is greater than the number of elementary particles in the known universe, then, given the variables of carbon content, alloy content, innumerable mechanical and thermal treatments, and invisible microconstituents, the forging of iron or steel produces a googolplex of microstructures, no two of which are alike. See kan of the ironmonger.

Gossans: Oxidized iron pyrite in the form of outcrops of decomposed rocks; a source of iron for early smelting furnaces.

Grain: In ferrous metallurgy, the consistent alignment of microstructural crystals, the patterns of which allow a blacksmith to judge the mechanical properties and quality of iron and steel. For example, grain size in cast iron is determined by the rate of cooling of the iron in molds, as well as its silicon content. High silicon content (2 - 3%) produces large graphite flakes and a coarse grain structure. See fracture, combined carbon, and microstructure.

Grain boundary kinetics: Phase transformation by grain boundary diffusion, also called diffusion creep, is a characteristic of non-superplasticity in metals, typified by the formation of ferrite-pearlite structures during the slow cooling of blister steel. Grain boundary sliding by dislocation creep is a characteristic of superplasticity in metals and is facilitated by fine grain size and lack of cavitation.

Grain direction: Expresses the uniformity or lack of uniformity of the microstructure of the iron or steel being forge welded. The ability of the smith to follow the strain patterns, expressed as grain direction, of the metal and incorporate them in the design of the implement to be manufactured is key to successful forge welding. "In a gun hammer, the strain is along the nose, across the finger and down the body of the cock" (Greener 1910, 246). One of the primary objectives of the faggoting, piling, and reforging of wrought iron into bar stock of the highest quality was the correct alignment of the grain direction of the siliceous fibers remaining in the wrought iron.

Grain growth: Increase in the grain size of crystals within the lattice structure of iron carbon alloys, which occurs when the metal is heated above the critical temperature and held at that temperature for a time sufficient for adjacent crystals to absorb each other.

Grain size: Along with homogeneity of carbon distribution, grain size determines the mechanical properties of steel, including the strength, toughness, and hardness of the crystal structure of iron and steel. Coarse-grained austenite increases hardenability in steel. In other cases, increasing grain size in polycrystalline structures decreases hardness. Decreasing grain size increases hardness, as in the dispersion of fine cementite particles during the formation of martensite. Austenite grain size coarsens with increasing heating temperature and influences hardenability by slowing the cooling rate of the interior of a piece of steel being quenched while the exterior becomes hard martensite. "Fine grained martensite usually transforms to fine-grained ferrite, pearlite, bainite or martensite, thereby providing increased strength and toughness" (Sinha 2003, 10.7).

Grain structure: Microstructural arrangements of the crystals in iron carbon alloys that determine their properties, i.e. coarse grained, fine grained, stressed (slip plane dislocations, as in martensite shearing). See stress fields, microstructure, lattice structure, and grain direction.

Graphene: A one atom thick sheet of pure carbon characterized by high conductivity, high electron mobility, and a free standing two dimensional crystal microstructure. "Graphene nanoribbions may prove generally capable of replacing silicon as a semiconductor in modern technology" (Wikipedia 2013). Graphene transistors are the wave of the future for communications technology. The recent discovery of the existence of graphene as a component of carbon nanotubes may unravel some mysteries about the microstructure of Damascus steel swords.

Graphite crucibles: Crucibles for casting steel made from a combination of clay, lead, and graphite. Developed by Joseph Dixon in 1850, the thermal efficiency of the lead (plumbago) in American-made crucibles, which allowed casting at high temperatures, played a role in the rise of American crucible steel production after 1860. See Stourbridge clay and cast steel 1.

Graphite flakes: Randomly occurring, bent, wrinkled, crystal scales of graphitic carbon, which weaken the structure of cast iron by creating spaces in its crystalline structure. See crystalline structure, graphitic carbon, and grey cast iron.

Graphitic carbon: Free graphite flakes, as in wrought and cast iron. The relative proportion of free graphitic carbon flakes and chemically combined carbon in cast iron determines its structure and physical properties, i.e. whether the iron is brittle or durable. The amount of the graphitic carbon is dependent on the cooling rate of the iron. The slower the cooling rate, as in sand molds, the higher the free graphite content. Free graphite weakens the iron but also increases its machinability. White cast iron is characterized by the predominance of chemically combined carbon, which is produced by rapid cooling, i.e. in iron molds. It is extremely durable but difficult to machine. See malleable cast iron, white cast iron, grey cast iron, and cooling rate.

Great migration (1629-1643): Migration of tens of thousands of Puritan dissenters to the Massachusetts Bay Colony, beginning with the Winthrop fleet of 1629, which transported hundreds of knowledgeable blacksmiths, bloomsmiths, shipsmiths, and shipwrights to colonial New England. This migration was the foundation for the rapidly expanding ironworking and shipbuilding economy of New England that was well underway by 1635.

Green sand mold: Mold used for casting grey cast iron, malleable cast iron, and other non-ferrous metals such as bronze, brass, and aluminum. Damp sand is rammed around a pattern and, when the pattern is removed, the sand mold is filled with molten metal. See casting and molder.

Grey cast iron: Cast iron containing free graphite in the form of graphite flakes as its principal form of carbon and having its combined carbon content not in excess of the eutectoid percentage. Slowly cooled cast iron is characterized by larger than normal acicular graphite crystals, which are very weak, making cast iron that is very brittle. Produced in its modern form in hotter, coke-fired blast furnaces, its higher silicon content inhibits cementite formation and makes the grey cast iron more difficult to decarburize. It can occur as a matrix of ferrite and pearlite or as a mixture of both and is the principal industrial form of cast iron. It is also used to make malleable cast iron by chilling and annealing. This process changes the large grey flakes of free graphite to a more finely divided pattern of graphite distribution, promoting the formation of graphite-austenite microstructures. Free graphite in grey cast iron that occurs as loose brittle flakes cuts through and separates grains of iron and makes the cast iron brittle rather than malleable resulting in low tensile strength and high thermal conductivity. The annealing of grey cast iron restores malleability. Chemically combined carbon prevents malleability in white cast iron, which is also annealed to make malleable cast iron. The addition of silicon and graphite stabilizing elements enhances the fluidity and durability of annealed cast iron. Modern grey cast iron is composed of approximately 4% carbon, 2% silicon, 1.5% manganese, 1% phosphorous, and 0.05% sulfur. The addition of chromium and vanadium facilitates the formation of carbides, producing fine grained, high strength cast iron (Pollack 1977). For some grey cast irons, nickel in a range of 4 – 15% may be added to control graphite flake size and enhance heat and corrosion resistance. *Tool Talks* (Stanley Tools 1937, 15) describes grey cast iron for use in the manufacture of Stanley planes. This entire article on grey iron casting is available on the Davistown Museum website as a tool information file. See cast irons, ferrite, graphite, malleable cast iron, pearlite, and white cast iron.

Grind: Removal of material by abrasive action; for edge tools, to wear down, polish, or sharpen using friction. Grinding by motorized machinery using vitrified bond wheels consisting of a mixture of emery and pottery slip clay replaced grinding utilizing sandstone wheels and hand files for most toolmaking functions beginning in 1878 (Lee 1995, 34). The use of a sandstone wheel mounted in a wood or metal frame continued in rural communities for sharpening axes until well into the 20th century.

Grindability index: A designation of the ability of a high speed tool steel cutting tool to be re-sharpened or otherwise altered due to the presence of alloy related hard metallic carbides.

Grindstone: Traditional tool for sharpening axes. As used in the 19th century ax-making industry prior to the invention of the vitrified bond wheel by Frank Norton of Worcester, MA, grindstones were 6 to 12 inches thick, 4 to 8 feet in diameter, and weighed 2000 to 4000 pounds (Lee 1995, 34). Smaller versions of ax grindstones mounted in a wood or metal frame are still commonly encountered in rural environments, in which stone thickness is typically 3 to 4 inches and diameter range is 18 to 36 inches. Most grindstones were traditionally water-cooled. Sandstone was the preferred material because it was the only stone to release particles at a controlled rate during the grinding process. See ax-making techniques and eyepin.

Gun barrel iron: Sheet wrought iron folded around a rod and then welded to form the gun barrel. See gunsmithing and Belgian iron.

Gunmetal bronze: Alloy containing 90% copper and 10% tin.

Gunsmithing: Making of guns by a gunsmith. Prior to the advent of the all cast steel gun barrel, gunsmithing usually involved the pattern-welding of piled soft iron and, occasionally, strips of silver or shear steel (refined German or blister steel). Apparently, soft wrought iron (<= 0.08% carbon content) was preferred to malleable iron (0.08 – 0.5% carbon content) due to its ductility and ease of workability. The exception to the generic production of guns from wrought iron as welded iron and steel are the exquisite all steel wheellock guns produced by German gunsmiths in the late 16th and early 17th centuries for the nobility, professional military officers, and royalty, examples of which are currently on display in the Arms and Armor section of New York's Metropolitan

Museum of Art. The exacting techniques used to forge this high quality steel remain unexplained in contemporary texts. See Belgian iron, pattern-welding, and wrought iron.

Gutterman: In the operation of a blast furnace, the person in charge of supervising activity in the lower region of the blast furnace, especially the efficient drainage of liquid cast iron from the bosh through the temp to the molds. In the case of blast furnaces producing pig iron rather than hollowware, the gutterman would direct the melted iron into the rows of sows, creating cast iron ingots with the appearance of pigs, hence the name pig iron. Most pig iron was remelted in cupola or other special purpose furnaces for its intended use. See temp, Bosch, blast furnace, cupola furnace, and molder.

H1 Stainless steel: A new stainless steel made in Japan that is "precipitation-hardened and contains nitrogen instead of carbon as the primary alloying element." Used for knife-making, "the people who have used it claim it is completely impervious to oxidation." It is composed of carbon: 0.15%, chromium: 14.00-16.00%, manganese: 2.00%, molybdenum: 0.50-1.50%, nickel: 6.00-8.00%, nitrogen: 0.10%, phosphorus: 0.04%, silicon: 3.00-4.50%, and sulfur: 0.03% (Erin Casson 2013, personal communication).

HRTEM: High-resolution form of the transmission electron microscope used in the early 21st century for the analysis of the microstructure of metals. A HRTEM has the ability, not previously available to archaeometallurgists, of analyzing structures of nanometer dimensions (billionths of a meter, as in the nanowires of cementite in Damascus swords – 10 - 20 nm in diameter, 200 nm in length) (Levin 2005). See microstructure and superplasticity.

Haft: Helve or handle of an ax or adz set up in a shaft hole at a right angle to the major axis of the tool. See eyepin.

Halstadt: First Iron Age culture in Europe, named after the salt-mining town in Austria, where Celtic metallurgists switched from making bronze to iron tools c. 750 BC. See *Hand Tools in History* Volume 6 and La Téne.

Hammersmith: Another name for the Saugus Ironworks (1646 to +/-1676), America's first integrated ironworks.

Hand tool: Instrument of manual operation.

Hard facing: The use of welding methods to add a coating, edge, or point on a metal, especially an alloy, to resist impact, corrosion, abrasion, or heat.

Hardenability: Ability of steel to be hardened. The following factors determine hardenability: carbon content, grain size, lattice structure, heating temperature, heating and cooling rates, and prior thermal treatments. "The depth to which steel can be hardened to martensite under stated conditions of cooling is called its hardenability" (Shrager 1949, 145). "A mechanical and retrospective expression of the processes and phenomena occurring in steel products on cooling from austenizing temperatures" (Cias n.d., 3). "The capacity of steel to avoid transformation to soft products with a lower and lower cooling rate regardless of just what maximum hardness it may achieve in the martensitic state" (Bain 1939, 46).

Hardening: The process that occurs during the heating and forge welding of iron (alpha iron) with a carbon content of 0.5% or greater due to sudden cooling (quenching) of the metal from a temperature at 910° C or above, which freezes the lattice structure of austenite (gamma iron) in a hard and brittle form, which is difficult to work unless softened by tempering or annealing. Hardness is thus a function of tempering temperature. Tempering relieves stress by allowing lattice rotation, changing the crystal structure of the steel. "If steel contains less than four tenths of one percent of carbon it has little or no hardening power under this treatment [sudden cooling]; but steel with six tenths of one percent or more of the element, has the wonderful property of being slightly malleable in the annealed state but extremely hard and brittle after this sudden cooling" (Spring 1917, 109). See annealing, carbon content of ferrous metals, recovery, and temper.

Hardness: "Resistance to plastic deformation by penetration, scratching or bending" (Shrager 1949, 367). The sudden increase in hardness following hardening increases tensile strength but reduces toughness and ductility, which are restored by tempering. See hardening.

Hardness formation effectiveness: Slowly cooling austenite to produce lamellar structures of pearlite produces steel with different mechanical properties than the production of spheroidized structures from quenched and tempered steel. The lamellar formations of slowly cooled austenite have greater tensile strength than rapidly quenched and tempered austenite, but the dispersion of the carbide spheroids of the quenched, then tempered martensite provides greater elasticity (plastic flow) than the pearlitic microstructures characteristic of slowly cooled blister steel (Bain 1939).

Hardness penetration: Depth of the hardened surface of quenched tool steel, often increased by the addition of alloys (Palmer 1937). Hardness penetration is a prerequisite for many modern machine designs, including aeronautical and aerospace applications and is facilitated and enhanced by the use of combinations of alloying elements, the most effective of which is molybdenum (Cias n.d.). See alloy steel and tool steel.

Hardness scale: Graphs of measurements made by pressing a steel ball (Brinel test), a diamond (Vickers test), or a diamond cone (Rockwell test) against a metal surface. The diameter of the resulting impression is expressed by the reporting unit of kg/mm^2. See oil temper and Rockwell hardness test.

Hearth: Area in the bottom of a furnace, in which the bloom of iron or molten pig iron is contained until removed by tongs or drained by tapping; often accompanied by liquid slag.

Heat engine: Machine that utilizes heat energy from biomass fuels, which is then converted to forceful motion (work), as, for example, in the form of steam under pressure. The internal combustion engine is another form of heat engine that uses combustion gases to do work. Heat engines either use biomass as fuel or the heat generated by nuclear energy to power steam turbines and other machinery. See biocatastrophe, industrial infrastructure, and prime mover.

Heat engine fuel shortage: The fundamental challenge of modern industrial society is to find alternative energy sources for its industrial heat engines. New age ironmongers will be smelting iron with renewable resources to make functional artifacts and art for creative economies in the post-apocalypse. See infrastructure collapse.

Heat treatment: Modern steelmaking techniques include five major forms of high treatment, many the results of centuries of ironworking. Thermal processing of ferrous materials includes hot forging, quenching, tempering, annealing, and a variety of additional heat treatment processes including, normalizing, spheroidization, and stress relieving, all of which change their microstructure and, therefore, their properties. Rapid cooling of austenized steel creates martensite, a high strength, easily fractured material, rather than ferrite, cementite, and/or pearlite microstructures. For edge tool production, martensite must be tempered in a range of 200° C to 500° C, precipitating and uniformly distributing iron carbide particles (cementite) to relieve brittleness. Quenching and tempering may involve modified techniques using two or more quenching media, such as oil, water, or brine, e.g. austempering and martempering. See annealing, austempering, cubic structure, martempering, temper, and martensite.

Helveman: Hammer man at a puddling furnace or finery.

Helve hammer: Traditional tool used by the bloomsmith in a water-powered bloomery forge to shape the loup of wrought iron and expel excess slag and other contaminants. First developed in Europe in late medieval forges in conjunction with the appearance of blast furnaces, the helve hammer was traditionally made of wood with an iron hammer head and was a component of most direct process bloomery forges, including those operated in the US from the early North American colonial period to the beginning of the 20th century. In the 19th century, lighter weight tilt hammers with smaller beams operating at higher speeds (400 strokes per minute) evolved from the more primitive helve hammer designs. See Gordon (1996, 93) for an excellent illustration of a helve hammer.

Hematite: "Ore mineral (Fe_2O_3) containing 70% iron" (Gordon 1996), i.e. ferric oxide, the most commonly mined ore in the U.S. Hematite is a common deposit at the bottom of marshy ponds in its hydrated form, i.e. limonite (bog iron), and is a common component in bedrocks of all ages.

High frequency induction furnace: Modern form of the electric furnace, in which a high frequency alternating current is induced via a water-cooled copper core, producing the heat for smelting the charge. They are used especially for high alloy steels, including high speed tool steel, and for specialty production, such as marine hardware. See furnace types.

High impact hardness: Resistance of steels to fracture and deformation.

High pressure non-condensing steam engine: Improved, more efficient version of

Watts' steam engine. Invented in 1790 by the American, Oliver Evans (1755 – 1819), it represented a major innovation in heat engine design and played an important role in the rapid industrialization of western economies and the rise of the factory system.

High-resolution transmission electron microscope: See HRTEM.

High speed tool steel: Alloy steel formerly made by the crucible process and more recently manufactured in electric furnaces. High speed tool steels are differentiated from alloy steels in the AISI classification system for steel as having a minimum alloy content above 4%. Commonly used alloys characterizing high speed tool steels include tungsten, vanadium, cobalt, chromium, manganese, and molybdenum. High speed steel-cutting tools retain their hardness when heated to a red heat, whereas carbon steel cutting tools lose their temper with cutting speeds above 20 to 30 linear feet per minute. Oil-hardened tungsten steel can cut up to 200 to 300 linear feet per minute on a milling machine or lathe without losing its temper. There are now hundreds of varieties of high speed tool steel alloys used in manufacturing processes. See alloy, *Appendix V*, and steel, types of.

High strength low alloy (HSLA): A descriptive term for a modern type of steel that provides better mechanical properties or greater resistance to corrosion than carbon steel. HSLA steels vary from other steels in that they are not made to meet a specific chemical composition but rather to specific mechanical properties (Wikipedia; March 2013).

History of Woodworking Tools: Definitive survey of the history of woodworking tools by W. L. Goodman (1964) first published in 1964 and reprinted several times. This text contains information about early tool forms (e.g. socket hole vs. shaft hole). Goodman discusses woodworking tools dating as early as 2540 BC Egypt and has important information on Roman and medieval forms, as well as illustrations from Moxon. Citing a Russian archaeometallurgist, Goodman, has this comment on 11th and 12th century carpenters' tools found in a Russian late medieval horde: "Of 22 axes tested, 7 had a steel tip welded over the iron base; in 7 others, the steel was inserted between the folded iron of the head. Seven others were solid steel and one of iron throughout. In most cases, there were signs of subsequent tempering, confined of course to the cutting edge" (Goodman 1964, Kolchin 1953).

Hollowware: Artifacts other than ordnance made at blast furnaces or cupola furnaces by casting pig iron directly into molds e.g. kettles, fire backs, stove plates, anvils, etc. The slitting mill cylinders used by the nail makers of Wareham, MA, and elsewhere were cast *in situ* at the blast furnaces of Carver (Murdock 1937). See grey cast iron and molders.

Hogging: Bulk material removed by high speed tool steel cutting tools.

Homogenous transformation: Phase change transformation in which precipitation occurs uniformly throughout the entire matrix, usually characterized by continuous enlargement of initially small fluctuations within the supersaturated or undercooled solid solution (Sinha 2003).

Hoop iron: Thin strips of malleable iron used to hold the staves of the coopers' cask in place; often made by the shipsmith, as well as the village blacksmith, from the North American Colonial Era to the mid-19th century.

Horimono: Decoration of Japanese swords by the addition of ornate carvings or grooves prior to the final polishing of the sword. See tsuchioki, yaki-ire, and yaki-modoshi

Horse power: One horse power is the lifting of 33,000 lbs. to a height of one foot in one minute. The maximum horse power achieved by the editor of this glossary in unloading hand tools from his truck after a buying trip to southern New England is approximately 1/40 horse power.

Horsing irons: Long-handled caulking irons used to seal the deck butts and garboard seams of wooden ships.

Hot- and warm-forging: Strategy used by early Iron Age blacksmiths to forge edge tools and swords by alternating the temperature of the steel being forged; especially used by Damascus swordmakers working with high carbon Wootz steel to produce the discontinuous patterns of the damascened sword by repetitive hammering. No metallurgical analysis is yet available to indicate if New England edge toolmakers of the late 18th to mid-19th centuries did or did not use this technique.

Hot blast: Developed in 1828 by Nielsen in England, cold air for the blast furnace was preheated in a separate oven already heated by the waste gasses of the furnace. The hot blast increased blast furnace temperature, slag fluidity, overall furnace efficiency, and reduced fuel consumption. In the United States, the development of the hot blast furnace encouraged the substitution of coal and coke for charcoal as a fuel, stimulating the construction of larger, more efficient blast furnaces. In England and continental Europe, the hot blast increased the efficiency of furnaces already using coke as a fuel. See blast furnace and coke.

Hot forge: To carburize raw iron in a carbon-rich environment. See carburize and enclosure.

Hot molding capacity: Essential characteristic of superplastic metals at high temperatures in order to form complex shapes, accurately replicating essential industrial molds, such as aircraft turbine blades. See mold and superplasticity.

Hot press: Technique of using dies to form self-welded molded steel or ceramic artifacts under high pressure and temperature from atomized powders of ore-alloy mixtures.

Hot set: Chisel with a handle used to cut iron bar stock when at a bright red-orange color.

Hot short: Disintegration of hot steel during forging due to its contamination with liquid iron sulfides from ore or fuel-derived sulfur. Iron sulfides are preferentially removed as slag constituents by manganese-containing flux, as occurs with the addition of spiegeleisen to cast iron or the natural occurrence of manganese in rock ore, eliminating brittleness in steel at high temperatures. "When [metal is] unworkable at a welding heat it is [said to be] *hot short*" (Brewington 1962, 15). See sulfur, phosphorus, and ferrosilicate.

Hot strength toughness: Ability of alloy steels to resist wear, abrasion, and brittleness.

Hot, round, and crowded: Also polluted, infected, genetically modified, hormonally disrupted, incontinent, feminized/masculinized, unemployed, uninsured, uninformed, manipulated, marginalized, indebted, disenfranchised, powerless – and angry: The tale of the unfortunate decline of western finance capitalism market economies in the decimated round-world commons of our giant integrated ecosystem (Brack 2010). See flat world.

Hot white heat: Temperature at which the breakdown of spheroidized carbide molecules in high carbon steel occurs ($< 1200°$ C); also, the normal temperature for forging low carbon steels and malleable iron due to their higher melting points. The dissolution of the cementite phase at white heat results in the embrittlement and weakening of the steel microstructure, which will then crumble upon hammering.

Hot-work die steels: Modern alloy steels containing chromium, tungsten, and molybdenum. These alloys are resistant to cracking and thermal shock during use in high temperature die-forging applications. Hot-work die steels may be further hardened by heating to a temperature of $1850°$ F, air- or oil-quenched, and then tempered. See alloy, deep hardening, quench, and temper.

Hot working range: Temperatures at which iron maintains plasticity at or above its crystallization temperature ($700 – 1250°$ C), allowing it to be reshaped into various tool forms. During hot working, including rolling or forging, the lattice structure of the metal is simultaneously deformed and reformed with a tendency to form more uniform grain structures. See eutectic, eutectoid, lattice structure, and critical temperature.

Hurricane Sandy: A powerful hurricane that devastated parts of New York City and New Jersey in late 2012. Cataclysmic climate change-derived infrastructure collapse anybody?

Hydrate: To become combined with water, as in bog iron, a hydrated form of limonite. See bog iron and hematite.

Hydraulic press: Press now used for manufacturing most drop-forged tools. The hydraulic press squeezes instead of hammers metal into shapes, e.g. hammer heads, ax heads, etc., utilizing two-piece dies often made of high carbon steel. Originally used for drop-forging heavier castings, e.g. railway car axles. See die casting, die driving, drop-forging, and die-sinking.

Hydrofracking: The recently expanded technology of deep well vertical, and then horizontal, drilling for gas and oil in underground geologic deposits such as the Marcellus Shale of eastern North America. Hydrofracking utilizes a complex mixture of toxic lubricants to facilitate extraction contaminates large volumes of water in underground aquifers, exacerbating the world water crisis. The prospect of the combustion of ever increasing amounts of greenhouse gas producing nonrenewable biomass fuels assures the prospect of cataclysmic climate change by accelerating global warming.

Hypereutectoid steel: Steel with carbon content above 0.83%, e.g. pearlite, with an excess of cementite at grain boundaries, which makes it harder and more brittle than hypoeutectoid steels.

Hypoeutectoid steel: Steel having less than 0.83% carbon content, i.e. less than the eutectoid composition, characterized by an excess of ferrite at grain boundaries, giving it the soft and ductile qualities of low carbon steel and malleable iron.

Ice cream: The eutectic (melting) point of ice cream is 32°. In a sense, it is similar to steel, an iron carbon alloy with a much higher melting temperature than ice cream. Steel, however, has a much wider range of melting temperatures due to its carbon content and also a wider range of temperature gradients between its solid and liquid form. In between the two temperatures, steel is mushy. Mushy ice cream would have a smaller temperature range between its solid and liquid forms. When frozen to 400° below zero, ice cream has something in common with martensite steel: it is hard and brittle and cannot be spooned easily and it needs to be tempered to soften it up a bit. Severely frozen ice cream does not lend itself to annealing; its re-crystallized lattice structure makes it unpleasant to ingest. A final observation on this unfortunate metaphor: Ben and Jerry's Chunky Monkey and Cherry Garcia are the equivalents of various tool steels to which alloys have been added, i.e. tungsten (cherries), manganese (chocolate chunks). These compare with soft serve Dairy Queen style ice cream, just the basics, as with low carbon steel produced by the Bessemer process, where all the interesting contaminants, some of which help make convivial edge tools, e.g. silicon, have been burned out of the steel (Illich, 1973). Soft serve can be made more palatable. Just add the hot fudge (manganese). See critical point, eutectic, liquidus, martenizing, and solidus.

Color	°F	°C
White	2200	1200
Light yellow	1975	1080
Lemon	1830	1000
Orange	1725	940
Dark orange	1680	890
Salmon	1550	840
Bright cherry	1450	790
Cherry	1375	745
Medium cherry	1275	690
Dark cherry	1175	635
Blood red	1075	580
Faint red	930	500

Table 3. Temperature colors (Palmer 1937, 129)

Iconography of tools: Depiction of tool forms throughout history including in art, catalogs, and reference books. The illustration of tool forms on wall drawings in Egyptian pyramids and shaft furnaces and blacksmith tools on Greek vases are early examples of the consistent tendency to illustrate tool forms in art, particularly in printed books after 1350. Also the photographic depiction of tools. See Phenomenology of tools, *Diderot's Encyclopedia*, Moxon, pattern books, History of woodworking tools, and *Tools Teach: An Iconography of American Hand Tools* (Brack 2013).

Impact strength: Measurement of the ability of a metal to withstand fracture under shock.

Impingement: The engulfing of carbide particles by growing austenite crystals after they have *first been* nucleated by an austenite crystal during the formation of austenite (Paxton 1967, 10-1).

Incandescent temperature colors: The colors used to judge the hardness and forge-ability of steel from the early Iron Age to the mid-19th

century, when instrumentation became available to measure high temperature in heated metals by measuring their expansion, e.g. pyrometers. The key color for the blacksmith quenching hot iron to make steel is the cherry red heat (1375° F/750° C), which is the minimum quenching temperature, or critical point, of steel. See critical point, edge tools, and temper.

Income disparities: The rapid growth in income disparities is a component of the evolution of a global market economy and its prime mover, the Information Technology Revolution. The extreme accumulation of wealth by the klepto-plutocracy (the 0.01%), the robust accumulation of wealth by the opulent (the famous 1%), and the modest but now declining wealth of the semi-affluent beneficiaries (the 10%) of America's now withering free enterprise system will not be sufficient to offset social unrest of the other 89% who suffer the consequences of income disparities and lack of public resources as the Age of Biocatastrophe unfolds.

India iron: Iron first produced in southern India circa 1100 BC, followed by the production of Wootz steel at a later undetermined date, but no later than 300 BC (Srinivasan n.d.). The origins of iron and steel production in India appear to derive from a combination of pre-Islamic Hindu and Islamic Indian metallurgical traditions.

Induction furnace: Form of the electric arc furnace in which the primary current is carried by a coil of copper tubing circling the shell of the furnace, which generates a strong secondary current in the charge of ore and flux, which is quickly heated by the resistance to the secondary current. In electric arc welding, the same principle is utilized, i.e. secondary currents result in the electric fusion of two pieces of metal being welded. After slag skimming, discreet quantities of alloys can be added to create a variety of alloy steels.

Induction hardening: Process for surface and localized hardening produced by the heating of steel placed in a high frequency field (9,000 – 300,000 Hz) in an electric arc furnace, where resistance to the creation of magnetic fields by eddy currents heats the work piece (Pollack 1977). See electro fusion and induction furnace.

Industrial infrastructure: Interwoven technological grids of pyrotechnic society in the age of global climate change, which include telecommunications, mass transportation, the electrical grid, energy production and distribution, industrial manufacturing, consumer product manufacturing and distribution, water resources and availability, oceanic fisheries, and terrestrial food production and distribution. Hyper-digital information and communication technologies increase the efficiency and productivity of the global industrial infrastructure but also make it more vulnerable to a predatory shadow banking network, cyber-terrorism, and natural events such as geomagnetic storms (solar flares) and extreme weather. See biocatastrophe.

Industrial metallurgy: With respect to ferrous metallurgy, the development of bulk process steelmaking technologies after 1870 characterized by the successful implementation of the Bessemer pneumatic and Siemens open-hearth processes.

Infrastructure collapse: Anomalous, unplanned, unpredictable consequences of the synergistic interaction of natural events (hurricanes, fire, earthquake, drought, sea level rise, cataclysmic climate change, antibiotic resistant and viral infections, and geomagnetic storms), and human activity (warfare, greenhouse gas emissions, resource and ecosystem depletion, overpopulation, and the global transport of ecotoxins) with the interwoven technological grids of a pyrotechnic society. The worst-case scenario for infrastructure collapse will result from the impact of global biocatastrophe on the interwoven technological grids of pyrotechnic society. See biocatastrophe, deforestation, and industrial infrastructure.

Ingot iron: Modern form of highly refined iron with a very low carbon content produced in open-hearth furnaces operating at very high temperatures, e.g. Armco iron, which has a carbon content of only 0.015% (Baumeister 1958, 612-3).

Ingot: Cooled steel and iron castings derived from a cast iron mold after smelting. After solidification, these castings are subsequently reheated for further mechanical and thermal treatment by forging, rolling, or other forms of processing. See iron ingot.

Integrated circuit: The prime mover of the Age of Information and Communication Technologies. An integrated circuit consists of a set of electronic circuits on one small plate now characterized by millions of photolithographic transistors per mm^2 of microchip. Digital nanotechnology produces huge gains in the efficiency and productivity of global consumer society. It is also the basis of the high speed electronic computerized trading systems of a predatory shadow banking network. Born-again ironmongers and back-to-the-land activists can, nonetheless, harness these technologies for their own sustainable economies, the risk of psychic numbing notwithstanding.

Integrated ironworks: Indirect process, multi-purpose ironworks including a blast furnace, refinery, chafery, and the multiple hearths of blacksmith/shipsmith forges. The Saugus Ironworks historic site is an excellent reconstruction of an early colonial integrated iron works.

Interdendritic shrinkage: Tendency of crystals in the microstructure of ferrous metals to freeze in irregular patterns, leaving interstitial spaces within the lattice structure of the metals. "… an alloy will rarely freeze with a smooth surface and even a pure metal will reveal partially formed crystals" (Smith 1960, 128). See alloy, crystal structure, lattice structure, and microstructure.

Internal combustion engine: Biomass devouring heat engine, which produces combustion gasses to propel private passenger cars, commercial trucks, aircraft, merchant shipping, and most railroads and mass transportation systems. See carbon footprint.

Internal stress: Residual stress within the lattice structure of a metal that remains after thermal and mechanical treatments; relieved by tempering or annealing.

Interrupted quenches: "An interrupted quench is one which is not carried through to the temperature at which transformation of austenite to martensite commences, but is interrupted at some higher temperature, in order to suppress transformation of austenite into pearlite and at the same time avoid formation of martensite. The quenching bath is kept at a stated temperature appropriate to the formation of the microstructure which is desired for the steel being treated" (Shrager 1949, 169). See austenite, bainite, martensite, pearlite, and temper.

Interstitial: Solid solutions of iron carbon alloys in which atoms (e.g. carbon) occurring within the empty spaces of the crystal lattice structure (e.g. ferrite) distort the structure.

Intuition: A critical component of the art of edge tool forging and swordsmithing prior to the understanding of the chemical basis of ferrous metallurgy. Based especially on the color and feel of the steel being forged and centuries of the empirical experience of swordsmithing, cutlery and edge toolmaking, the collective experience and secret formulas derived from the intuitive working of hot iron were called "kan" by the Japanese swordsmith (Miyairi). See kan of the ironmonger and rule of thumb.

Iron: "A heavy malleable ductile magnetic silver-white metallic element that readily rusts in moist air, occurs native in meteorites and combined in most igneous rocks, is the most used of metals, and is vital to biological processes" (Merriam-Webster Dictionary online 2008). Iron is the fourth most common element composing about 5% of the earth's crust. Iron has a melting point of 2793° F (1538° C) and is difficult to smelt in its pure form, but when carburized (combined with carbon), it undergoes microstructural changes, which increase its hardness. Malleable iron (0.02 – 0.5 % cc) becomes steel when its carbon content range is 0.5 – 2.0%. If its carbon content reaches a range of 2.0 – 4.0%, it becomes cast iron with a lower melting temperature of 2075° F (1149° C). Cast iron has a wide variety of forms depending on variations in heat treatment and alloy content. See carbon content of iron and steel, cast iron, malleable iron, and steel.

Iron Act: Act passed in 1750 by the English Parliament that prohibited the manufacture of iron and steel tools, hardware, and artifacts in the American colonies. In part, this act was a response to the Swedish export tax (1750) on charcoal iron, which resulted in an increase in the importing of bar iron from the colonies to England after 1750. This Act was never successfully implemented, and a robust North American colonial era iron industry had already been established at the end of Queen Anne's War and the Treaty of Utrecht in 1713 (Bining 1933). The Iron Act played a key role in stimulating colonial resentment of British rule, helping to lay the foundation for the coming revolution. See bar iron.

Iron carbide: Alloy of strongly bonded iron and carbon atoms with various patterns of carbon distribution; also, specifically cementite (Fe_3C). The homogeneity or heterogeneity of the carbon distribution pattern in iron tools, for example, determines the characteristics of the tools as cast iron, low carbon steel, wrought iron, or crucible steel. Iron carbide is not brittle at high temperatures. Rolling and forging can successfully break up iron carbide networks, especially in high carbon steel being worked at temperatures between 650° and 750° C, at which point the network is transformed into coarse iron carbide particles visible to the naked eye as layered structures, as exemplified by the damascened patterns on pattern-welded Wootz steel swords. See carbon content, crucible steel, cementite, cast iron, Damascus sword, Wootz steel, and wrought iron.

Iron carbide networks: Essential components of the microstructure of ferrous metals at certain temperatures. The formation of brittle iron carbide networks in high carbon steels > 0.8% cc as a result of undercooling (interrupted cooling) or quenching (rapid cooling to room temperature) is responsible for their low ductility and fracture vulnerability at room temperature, in contrast to the plasticity of carbide networks formed at high temperatures. The rate of cooling and undercooling determines the grain size of the microstructures formed by the iron carbide networks and thus influences the mechanical properties of the steel being produced. See carbon content of iron, microstructure, and microstructure variable.

Iron-carbon constitutional diagram: "A constitutional diagram each point of which represents the composition of steel or cast iron that is in equilibrium and that contains only iron and carbon" (Shrager 1949, 368). See *Appendix VI*.

Iron cored: One of three ax-making techniques, known as the overcoat method, where steel strips are wrapped and welded around an iron core, then forged into the desired shape. While not listed in Pleiner's (1980) sketch of knife-making strategies, this contemporary ax-making technique may represent an additional early Iron Age strategy for making knives and possibly edge tools. See ax-making techniques.

Iron-fuel contact: A perennial problem in older furnace designs prior to the development of the reverbatory furnace by Henry Cort in 1784, which eliminated fuel to ore contact. Earlier furnaces decarburizing cast iron or bloomeries producing wrought or malleable iron ran the risk of recarburizing their iron by oxidation, as well as of contamination of the iron by contact with unwanted products in the fuel, such as sulfur. See ferrosilicate, finery, oxidation, puddling furnace, and reverbatory furnace.

Iron ingot: Common form of reprocessed cast iron, often produced in cupola furnaces as the raw material for foundries and other special purpose furnaces. See malleable cast iron and foundry. See ingot.

Iron mill: Forges, often rolling mills, that reprocessed the iron ingots derived from blast furnaces and iron muck bars (bar stock) from bloomeries and puddling furnaces for special purpose applications. This term applies to iron production before the era of bulk process steel. See blast furnace, bloomery, and ingot.

Iron ore: Ore containing any of the iron oxide minerals magnetite (Fe_3O_4), hematite (Fe_2O_3), or siderite ($FeCO_3$), sometimes in various combinations. Iron ore also occurs as the hydrated iron oxides goethite and limonite, as in the bog iron deposits of southeastern MA.

Iron oxide: There are 16 known forms of iron oxide. In the smelting process, the most important is ferrous oxide, a highly volatile form of iron oxide that easily ignites and provides the spectacular fireworks displays seen in bulk process steel production. Iron oxide is produced by the oxidation of iron through natural processes, as well as by smelting. Ferrous oxide plays a major role in the decarburization of cast iron by facilitating the reduction process and the removal of carbon. Another important form of iron oxide is the iron ore magnetite (Fe_3O_4), which is formed when iron corrodes after contact with water. See carbon content, cast iron, decarburization, iron ore, oxidization, and slag.

Iron Road: Famous trading route used from the early Iron Age to the late medieval era, via which currency bars smelted in Austria (Styria and Carinthia, also called Noricum) were transported by wagon north to Linz on the Danube River and then upstream to portages linking Austria with the forges and swordsmiths of the lower Rhine region. This latter area was particularly active as the Merovingian swordmaking center of the early Frankish empires of the region (650 – 750 AD) and later during the German Renaissance. See Celtic metallurgy, Halstadt, La Téne, and Spathas.

Iron sand smelting process: Process which makes use of self-fluxing, black sands from the south shore of the Black Sea to produce iron and nickel steel, c. 1900 BC. This was the first documented industrial production of iron and steel tools, which, ironically, occurred at the height of the Bronze Age (Wertime 1980). Iron sands are also the source of iron smelted by Japanese samurai swordmakers. Finely granulated iron sands are a key link between the intentional use of rock ores and the late 20th century development of powdered ferrous metallurgy, an industry which is dependent upon the availability of finely granulated powdered iron. See Chalybean steel, flux, samurai swords, satetsu, and nickel steel.

Iron silicate: Along with ferrite, the principal component of wrought iron. It occurs as glass-like slag and is present in wrought iron in amounts that vary from 1 to 3% by weight. It is distributed in the wrought iron as highly visible threads or fibers, which run in the direction of rolling (mechanical processing). "In well made wrought iron there may be 250,000 or more of these glass-like slag fibers to each cross-sectional square inch. The slag content occupies a considerably greater volume than a percentage by weight would indicate, because the specific gravity of the slag is much lower than that of the iron based metal" (Aston 1939, 2-3). The iron silicate content of wrought iron provides unique qualities, especially resistance to corrosion and fatigue. The iron silicate content of early forge welded hand tools explains their resistance to rusting in comparison to modern drop-forged low carbon steel tools, which rust quickly when exposed to moisture. See drop-forging and ferrite.

Iron-smelting furnace: Direct process shaft or bowl furnace used by bloomsmiths from the early Iron Age until 1900 to produce wrought and malleable iron. Most such iron was reprocessed by fineries and shaped by chaferies and rolling mills into forms compatible with its intended use. This term could also apply, in the generic sense, to blast furnaces, in contrast to melting furnaces, such as the cupola or reverbatory furnace. See furnace types.

Ironmonger: Ironworkers following the trades of the bloomsmith, shipsmith, edge toolmaker, cutler, farrier, or community blacksmith. Ironmongers usually worked in darkened environments, often at night, where the lack of sunlight allowed them to make the critical judgment of whether the iron or steel tools they were forging or repairing were ready for quenching, tempering, or other thermo-mechanical treatment, as shown by the cherry red (1375° F) or other colors of iron and steel heated to appropriate temperatures. For further commentary on ironmongers, see the *On Ironmongers* preface. See born-again ironmonger, incandescent temperature colors, and kan of the ironmonger.

Isometric crystal forms of iron: Temperature dependent allotropic forms of iron, which include alpha iron (ferrite) < 1670° F (910° C), gamma iron (austenite) 1670 - 2540° F (910 - 1393° C), and delta ferrite > 2540° F (1393° C). The addition of alloying elements other than carbon can change the temperature range of phase transformation.

Isothermal transformation diagram: Time temperature transformation (ttt) curves in the form of diagrams that graphically illustrate the changes that occur when steel is cooled at various rates. The diagram for eutectoid steel (Shrager 1949, 143) illustrates the phenomenon that, at relative low temperatures (+/- 1000° F), the transformation of austenite into martensite can take place in a matter of seconds. The first isothermal transformation diagrams were formulated by Davenport and Bain in 1930. See critical temperature and *Appendix VI*.

Isotropic: Materials having identical properties in all crystallographic formations, i.e. their elastic forms are alike in all directions. See anisotropic and crystal structure.

Japanese samurai swords: Japanese swords are traditionally pattern-welded matrixes of layered iron and steel, re-piled and reforged in elaborate rituals that often result in hundreds of thousands or millions of steel strata. The heyday of samurai sword production was the Kamakura era (1185 - 1392). Japanese swords had two basic forms: banded steel jackets around a softer steel core and hard steel inserted between softer steel plates. Smith (1960) has an extended discussion of Japanese swordmaking techniques in *A History of Metallography*. National Public Television's NOVA (October 2007) had an excellent program on direct process smelting of the steel used for the samurai sword, including a vivid depiction of the use of a breakdown furnace to produce the bloom of raw steel. See natural steel and breakdown furnace.

Jominy end quench test: Standard test for determining the hardness of alloy steels after quenching. A piece of machined steel 1" in diameter by 3 7/8" long is heated to a quenching temperature, which is held for 20 minutes. Gradients in the cooling rate of the steel rod produce associated gradients of hardness. See hardenability and tool steel.

Juanita charcoal iron: Highest quality direct process smelted wrought and malleable iron bar stock produced in Pennsylvania in the 19th century; highly valued by Pittsburgh producers of cast steel. See cast iron.

Kan of the ironmonger: Combination of ritual and empirical experience of the Japanese swordsmith; a methodology of swordsmithing that was beyond words or written language. Swordsmiths had their own individual technique for the forging of the sword and learned their art under the tutelage of older swordsmiths. This intuitive rule of thumb understanding of the mysteries of forging hot iron was also the key to the successful production of Damascus, Viking, and Merovingian swords. Intuition and empirical experience were also the keys to the success of New England's edge toolmakers and sword cutlers in the decades before science provided its academic explanations of phase transformation kinetics. The kan of the ironmonger was intuitive in nature, a series of moments of epiphany, a technical finesse beyond written language learned by years of the trials, errors, and observations of the apprentice ironmonger working under the tutelage of master ironmongers, who themselves worked within the legacy of millennia of hot iron forging. See yaki-ire and yaki-modoshi.

Kentledge: Hardened blocks of pig iron used as ballast in coasting vessels and then sometimes sold to fineries and foundries for further processing. See iron ingots and pig iron.

Kera: Raw steel bloom produced in breakdown furnaces and used in Japanese samurai sword production. See furnace types and Japanese samurai sword.

Keyword index for cascading Industrial Revolutions: Terminology that summarizes key components of the cascading Industrial Revolutions (IR) of pyrotechnic industrial society— IRI: steam (>1750), IRII: machines (> 1840), IRIII: steel (>1875), IRIV: petrochemicals (>1900), IRV: science (>1945), IRVI: fiberoptics (>1995). The Age of Biocatastrophe (>2011): the struggle for survival in the context of gross income inequality. See cascading Industrial Revolutions.

Kill: "To remove dissolved gas in liquid steel so that it will not boil while solidifying" (Gordon 1996, 309); a key step in the successful production of high quality cast steel. See crucible steel.

Killed steel: Steel held in the crucible until no more gas is emitted and the surface appears "quiet;" a process that prevents blow holes and other structural anomalies (Spring 1917). When the steel in the crucible solidifies, uneven cooling causes shrinkage, which then produces a central pipe in the crucible casting. See dozzle and pipe.

Kiln reduction process: Modern process for the smelting of sponge iron in a gas-fired rotating kiln furnace, which is typically 150 feet long, 7.5 feet in internal diameter with a fire brick wall about 8 inches thick. In the kiln reduction process, the iron being smelted is moved slowly through the kiln by a conveyor and then dropped into a cooling kiln located below the main kiln. Large quantities of sponge or wrought iron can be produced by this method (Pollack 1977).

Kinetics: Study of the motion of bodies and the forces acting on them; with respect to ferrous metallurgy, the study of the changes in the allotropic forms and microstructural characteristics of iron alloys due to mechanical and thermal treatments and events. See phase and phase transformation.

Klepto-plutocracy: An elite class of the top world income earners who maintain economic and political control of a globalized market economy. In the US, the one hundredth of one percent includes a minority of plutocrats, such as Bill Gates, who redistribute much of their wealth for worthwhile social causes. The majority of the 0.01% are klepto-plutocrats who obtain their wealth by the manipulation of banking, stocks, and investment funds (e.g. the shadow banking network) or by the accumulation of wealth by utilizing tax loopholes (e.g. the carried interest rate) and off-shore tax havens. America's klepto-plutocracy is made possible by lobbyists who can subvert congressional efforts to ameliorate the federal tax codes, redirect sectarian political factions, such as tea party crypto-fascists, and an Information Technology Revolution that promotes computerized trading, media propaganda, massive structural unemployment, and the psychic numbness of flat world technology. The resulting rapidly increasing income disparities, political polarization and paralysis, and social inequities and unrest provides the context of and challenges for the sustainable economies and social networks of the future. In the Age of Klepto-plutocracy, finesse in the use of hand tools by ironmongers and other artisans will provide the basis for alternative lifestyles and sustainable economies. The dynamics of the Age of Round World Biocatastrophe will also impact the lifestyles of the klepto-plutocracy and techno-elite whose innovations and ingenuity are the prime movers of their wealth, as well as the one hope of mitigating the impact of the unfolding Age of Biocatastrophe. See biocatastrophe, flat world, income disparities, and shadow banking network.

Knife forms: There are five basic forms of knives, as described by Tylecote (1987, 269 after Pleiner 1969), which are derived from his excavations of central European archeological sites over a period of several decades. These knife forms existed more or less simultaneously in eastern, western, central, and European sites from the beginning of the early Iron Age (Halstadt +/- 700 BC) until the late medieval period (1000-1500 AD). The forms are: *all iron* knives, knives with *carburized* cutting edges, raw or natural *steel* knives, *sandwich pattern* knives (the iron components of the knife are folded over the steel cutting edge, the earliest known variation of welded steel construction), and *pattern-welded* knives, where alternating layers of steel and sheet iron are welded together. The last appears to be the most common knife form in archeological sites. See history of woodworking tools and pattern-welding.

Knife makers' steels: Traditionally, most knives and swords were pattern-welded combinations of steel and wrought iron. Until the mid-19th century the chemistry and alloy content of knife makers' steels were unknown. *Appendix VII* reproduces Latham's (1973) listing of popular knife makers' steels. The carbon content of these steels generally ranges from 0.85 to 1.5% cc. The wide variation in alloy content of these steels illustrates the complexity and diversity of the alloy content of steels used by knife and swordmakers, as well as edge toolmakers. The high chromium and molybdenum content

94

of many of these steels illustrates the important contemporary role of alloys as additives to knife makers' steels. For knife and sword makers working before 1850, the presence of significant alloys, such as manganese and silicon, may have been known through practical "rule of thumb" use of ores containing these compounds. After the development of modern chemistry in the late 19th century, alloys with precise combinations of additives were produced for specific types of knives, cutlery, and edge tools.

Knobbler: Bar maker at a finery who made anconies. The name derives from the fact that most anconies had a knob at the end to facilitate ease of movement. See ancony.

Lamellae: With respect to the crystal structure of steel, the plate-like structures that characterize pearlite. During the slow cooling of austenite, identical pairs of lamellae are formed by the process of twinning, creating pearlite, a form of steel with a more distorted crystal structure than that contained in tempered martensite. See austenite, bainite, cementite, laminae, martensite, Neumann bands, pearlite, and twinning.

Lamellar family of structures: The unique lath-like microstructure formed by the slow cooling of austenite. If heated to 1400° F (760° C), austenite with a carbon content ranging from 0.7 - 0.9% consists of polyhedral grains, which, when slowly cooled, revert to lamellar structures of ferrite and carbide (pearlite). Interrupted cooling results in the wide variety of lamellar structures called bainite. Increasing the rates of cooling increases the hardness of the steels formed while lowering the temperature of phase transformation (Bain 1939, 25). A higher carbon content results in a change in the fine-grained lamellar structure to increasing coarseness. During rapid cooling, lamellar transformations are halted at 1000° F (550° C) until the temperature falls below 100° C (200° F), at which time martensite is formed (Bain 1939, 25-8). See bainite and pearlite.

Lamellar pearlite: "Pearlite crystals covered by a thick layer of cementite" (Shrager 1949).

Laminae: With respect to gunsmithing and swordmaking, the patterns resulting from the piling and forging of alternating layers of steel and iron strips and plates, which form the composition of the pattern-welded gun barrel or sword. See damascene, gunsmithing, piling, pattern-welding, and gun barrel iron.

Laminar cementite: Lath-like formation of slowly cooled cementite. Another variation in what Bain (1939) called the "lamellar family of structures." See Levin (2005).

Laminated structure: Alternate layers of different allotropic forms of iron, such as the alternating layers of ferrite and cementite that constitute pearlite.

Laminating: Early Iron Age technique of knife-making characterized by pattern-welded layers of sheet iron or sheet steel with varying carbon content, as exemplified by Egyptian knives dating from 900 – 800 BC, as well as an adz from the same period. Laminating is a frequently used technique for steeling edge tools by the lamination of steel on the underside of the tool. See Wertime (1980, 120), knife forms, and pattern-welding.

Lap riveting: Riveting of overlapped sheets of iron and steel in boiler-making and iron and steel shipbuilding.

Lapping machine: Machine with cast iron disks having lapping compounds imbedded in the disk; used especially for fine surface finishes on the back of edge tools prior to final honing.

Lap weld: To weld pipe so that the skelped pipe edges are beveled, overlapped, and welded together; a variation of scarf welding. See scarf weld and skelp.

La Téne: Village and ritual site on the edge of Lake Neuchâtel in Switzerland. The tools uncovered at La Téne reflect both a western movement of Celtiberic metallurgists and the evolution of a more war-like culture than Halstadt. Iron axles made their first appearance in Europe in warrior burials at La Téne. See Halstadt and natural steel.

Lattice bending: Bending that results from the random application of stress on the lattice structure of metals, producing distortion patterns that lack crystallographic regularity (Barrett 1943, 307). See plastic deformation, cubic structure, and space lattice.

Lattice reorientation: Lattice reorientation of the crystal structure, of iron for example, occurs by three processes: uniform rotation of the crystals, bending of the lattices between planes of slip, and division of crystals into deformation bands "within which the lattice rotates in different directions" (Barrett 1943, 347-48). Heating and cooling are the principal causes of rapid lattice reorientation (slip); cold working is the principal cause of slow lattice reorientation (creep). See plastic deformation, slip, and creep.

Lattice structure: Crystal structure of the unit cells of metals. See allotropic forms of iron, crystallography of ferrous metals, cubic structure, unit cell, lattice reorientation, plastic deformation, and space lattice.

Ledeburite: Cast iron containing 4.3% carbon in the form of the eutectic mixture of austenite and cementite. The austenite in ledeburite may transform to ferrite and cementite during cooling.

Leg vise: Special form of vise characterized by a 3" to 8" wide jaw taper mounted on an iron leg with a bar spring mechanism. The presence of the leg vise, anvil, and associated hand tools characterized blacksmith forges and farm workshops where toolmaking and repair occurred. Leg vises can be traced back to Roman smithies and are illustrated in Diderot's *Encyclopedia* ([1751-75] 1959) and in English pattern books. Most leg vises recovered from New England workshops appear to be domestically produced and are often hand-forged, wrought steel, often one-of-a-kind creations of local workshops.

Leonard, James: An English ironmonger who migrated to the Massachusetts Bay Colony during the great migration and worked briefly at the Braintree Ironworks (Furnace Brook) before moving to Taunton, Massachusetts, and establishing his forge on the Two Mile River in 1652. With his brother, Henry, his son, Nathan, and many other members of his family, James was instrumental in establishing a vast network of bog-iron-smelting furnaces and forges from Rowley in northeastern Massachusetts to New

Haven, Connecticut. Many of these forges were still operating in the first few decades of the 19th century. The Leonard clan played a major role in supplying New England shipsmiths and toolmakers with domestically produced direct process smelted iron bar stock.

Lime as flux: Lime was used extensively to remove phosphorous from iron. By 1878, Sidney Gilchrest Thomas had elucidated the chemical basis for its usefulness as a slag constituent and as a furnace lining. See acid process, basic steel, flux, and limestone.

Lime boil: Violent reaction occurring during the basic open-hearth process when carbon dioxide gas is released from the limestone on the hearth floor. See furnace types and hearth.

Limestone: Stone used as a flux in iron-smelting after 1600; used specifically as a flux for high phosphorous iron ores after 1878. The calcium content of limestone greatly reduced the iron content of blast furnace slag, allowing recovery of 90% or more of the iron being smelted. See acid process, basic process, and lime as flux.

Limonite: Bog iron; also "all forms of hydrous sesqui-oxides of iron" (OED 1975, 1629). "Impure form of goethite containing a variable amount of water" (Gordon 1996, 309). Limonite is mined from bogs and heated to evaporate water, a principal source of iron for bloomeries and blast furnaces in the American colonies, especially in southeastern Massachusetts and the pine barrens of New Jersey. See bloomery and bog iron.

Liquidus: "The upper curve in a constitutional diagram, which is the locus of temperatures at which each alloy starts to solidify" (Shrager 1949, 370). See Barraclough's constitutional diagram in *Appendix VI*.

London spring steel: A common trademark on Henry Disston's highest quality hand saws, London spring steel is cast steel rolled into sheets. Disston used English cast steel to manufacture many of his saws until the 1880s, even though excellent cast steel was being produced in Pittsburg, PA after 1850. See spring steel.

Lost wax process: Ancient method of casting that utilizes wax patterns for molds, which are then destroyed by the casting process; particularly useful for precision casting, such as in jewelry making. The use of the lost wax process has been widely adopted by contemporary jewelers and artisans. See cire perdu.

Loup: Synonymous with "bloom," as in the "bloom" or "loup" of iron produced in the smelting process. See bloom, bloomery, direct process, and reduction.

Low carbon cast iron: Cast iron produced in air furnaces as white cast iron for the manufacture of malleable cast iron; contains about 20% less carbon than grey cast iron. See cast iron and grey cast iron.

Low ductility of high carbon steel: A characteristic of high carbon steel that has been cooled from high temperatures to intermediate temperatures. Its low ductility is a result of the formation of brittle iron carbide networks, as in the case of martensite shearing. At room temperatures, such high carbon steel is subject to cracking due to embrittlement before annealing or other thermal treatments reduce microstructural stresses. See carbon content of iron, martensite, shear steel, and sheering stress.

Machinability: The ease with which alloy steel and malleableized cast iron may be machined. Water-hardened alloy steels and malleableized cast iron have the best machinability characteristics.

Machine: Apparatus for applying mechanical power, consisting of a number of mechanical parts, each having a definite function. Machines are often powered by heat engines. See simple machines, turbine, and heat engine.

Machine steel: A modern form of "low carbon steel," usually with 0.08 - 0.2% carbon content but without the slag content of malleable iron; frequently drop-forged in dies to be made into hand tools other than edge tools; often subject to additional carburization, usually by case hardening. See carbon content of ferrous materials, low carbon steel, malleable iron, and case hardening.

Macro-etch: To treat the surface of a sword, for example, by a reagent, for the purpose of determining the quality of metal by analysis of the flow patterns that reveal the structural differences characteristic of the iron and steel used to fabricate the sword by pattern-welding. See Damascene, Japanese sword, and reagent.

Magdalenberg: Important early Iron Age metalworking site in Austria, located in Carinthia, near Styria and Erzberg (Ore Mountain); the location of many shaft furnaces and the probable production site of malleable iron and natural steel tools and weapons.

Magnetite: "Magnetic mineral (Fe_3O_4) containing 72% iron; the active ingredient in 'lodestone'" (Gordon 1996, 310). While hematite (Fe_2O_3) is the world's most common form of iron ore, magnetite is the primary source of tool steel for edge tools and industrial alloy steel products of all kinds. Among the world's most important historic sources of magnetite are the low-sulfur, low-phosphorus rock ores of northern Sweden, which were used to supply the steel manufactories of Sheffield. Swedish magnetite-derived wrought iron bar stock was an important transatlantic trade commodity and was imported to the United States in large quantities throughout the 18th and 19th centuries for American edge toolmakers.

Mainspring: Apparatus that replaced weights as a method of powering clocks. They first appeared in Germany in 1550 and became widely used in the florescence of German clock-making from 1550 to 1634 and in England and Holland after 1650, when the German industry began to decline. The search for higher quality steel for mainspring production led Benjamin Huntsman to readapt crucible steel manufacturing techniques for his Sheffield clock and watch spring business in 1742. See cast steel 1).

Male die: Upper die, also called a "force," made of steel, cast iron, or copper by a drop press in a machine shop or by a toolmaker, rather than by a die-sinker. A heated pattern blank is used to shape the male die by the drop press. The die is then hardened and tempered, making a reproduction of the pattern model. See female die.

Malignant tool dependency (MTD): Generally progressive condition ultimately linked to universal acquisition disorder, familiarly known as the collector's disease. As with other degenerative obsessive-compulsive conditions, MTD can deepen to the extent that the gathering of tools far exceeds need, either for use or profit. First state MTD can begin before formal schooling and may lead to a morbid late stage that can create a broadly negative presence in the subject's life (hoarding). Symptoms are numerous and varied, often well hidden behind an apparently rational barricade. Destructive patterns can continue long after corrective resolutions are made. There is no permanent cure (McLaughlin, David 2008, Liberty Salvage Co., personal communication).

Malleability: Relative ability of a material to be permanently deformed by rolling, casting, forging, extruding, etc. without fracturing and without developing increased resistance as it is deformed. Malleability usually increases when temperature increases. Wrought iron, gold, silver, aluminum, copper, and magnesium are malleable metals. See ductility, plastic deformation, and malleable iron.

Malleable casting: Malleable cast iron having some of the properties of wrought iron; derived from annealed cast iron by a process that involves the cleaning of the cast iron by pickling, followed by its stacking within the annealing oven after being covered with millscale or other forms of iron oxide. Rapid heating is followed by very slow cooling, which produces an iron that is resistant to damage by sudden shock and often machinable. See annealing, casting, and malleable cast iron.

Malleable cast iron: Cast iron, characterized by the predominance of free carbon in the form of graphite flakes made by the rapid cooling of white cast iron or from heat-treated grey cast iron. The most common and commercially useful form of cast iron because it is durable, strong, shock resistant, malleable, and machinable, its fracture and color express its metallurgical composition. Malleable cast iron is made in two forms: whiteheart and blackheart, which are differentiated by the processes used to anneal them and by their chemical composition (Spring 1917). White cast iron containing chemically bound graphite is unsuitable for making malleable cast iron until it is melted and annealed, freeing the bound graphite in the form of temper carbon, which softens the microstructure of the cast iron, allowing it to withstand distortion without breaking. Malleable cast iron is also produced in cupola air furnaces as pearlitic malleable cast iron, particularly useful for drop-forging hand tools, where shorter annealing times result in greater strength and wear resistance. Knight (1877) provides this definition: "Iron cast from the pig into any desired shape, and afterwards rendered malleable, or partially so, by annealing. A great variety of articles such as bridle-bits, snuffers, parts of locks, various forms of builders' and domestic hardware, some kinds of culinary and other vessels, pokers, tongs, and numerous other things can thus be produced in a more economical and correct manner than they could be by forging. Many of these are subsequently case hardened and polished. *The art of softening cast iron and working it, and afterward hardening it again.* [italics added] ...The inventor of the process of rendering articles of cast-iron malleable was Samuel Lucas of Sheffield, by whom it was patented in 1804" (Knight 1877, 1376-77). First produced commercially in the United States by Seth Boyden in Newark, NJ, in 1831, with various alloy combinations and heat treatments, malleable cast iron has been used to manufacture an endless variety of hardware and utensils. Writing in France in 1722, R.A.F. de Réaumur also describes this process. The widespread production of malleable cast iron tools in the 3rd and 4th century BC in China has also been noted (Needham 1958, Barraclough 1984). See annealing, grey cast iron, and temper carbon.

Malleable iron: Nineteenth century term for iron having a slightly higher carbon content (0.08 - 0.5% carbon) than wrought iron (< 0.08% carbon content), now commonly referred to as "low carbon steel." However, unlike low carbon steel, malleable iron contains variable amounts of silicon slag (more if smelted by a bloomsmith, less if refined in a puddling furnace from decarburized cast iron), as well as manganese, phosphorous, and sulfur. Silicon, manganese, and phosphorous can add beneficial qualities in tools being produced from malleable iron. Silicon enhances the ductility and durability of malleable iron, and manganese enhances strength. In small quantities, phosphorous increases hardness and has a useful role in nail making. Sulfur has no useful function in any metallurgical context. Malleable iron will harden slightly when suddenly cooled. Siliceous slag-bearing malleable iron was produced as a "bloom" or "loup" in direct process bloomeries. Iron with low slag content was produced in indirect process puddling furnaces from pig iron. Malleable iron containing significant levels of silicon is not suitable for production of tool steel alloys. The silicon content of malleable iron played an important, but unrecognized, role in the forge welding of hand tools, including edge tools, in the era preceding bulk process steel and machine drop-forging. The confusion between very low carbon wrought iron and relatively low carbon malleable iron continues, the latter being frequently referred to as wrought iron, while having the same carbon content as modern "low carbon steel." See carbon content of iron, Bessemer process, mild steel, and low carbon steel.

Malleableizing: Annealing of white cast iron for the purpose of changing chemically combined carbon to temper carbon, creating malleable cast iron that is less brittle, more durable, and more machinable than gray or white cast irons. See malleable cast iron and temper carbon.

Manganese: Metallic element Mn, often appearing in its oxide form in ores. Manganese is an important constituent in steel alloys (+/- 0.2%). It dissolves in ferrite, hardens and toughens it, serves as a deoxidizer, increases elasticity, and causes carbon to stay in combined forms as hardened steel. Manganese retards austenite transformation, neutralizes phosphorous, and also makes iron non-magnetic. It makes tool steel easier to hot roll or forge and increases the penetration of hardness during quenching. Water quenching of manganese alloy steels tends to make them too brittle for most applications. Historically, the manganese content of iron ore played an important role in natural steel production by lowering the melting temperature of slag, thus facilitating carbon uptake in a bloom of iron. The manganese would become part of the flux, preferentially combining with contaminants such as sulfur and phosphorous, which were removed when the slag was drained off of the bloom. As a constituent of cast iron, manganese played a similar role in the production of German steel. When the cast iron was fined, the manganese preferentially absorbed the sulfur, which was removed as slag. Uniformity of carbon distribution was facilitated, but only in fined cast iron containing the manganese. In the 19th century, the chemical basis of the role of manganese became known, and the addition of ferromanganese to the smelting process was the key to the success of Bessemer steel. It is also used in the modern production of rifle barrels with tough hardened cores. Siderite is the only naturally occurring iron ore containing manganese; notable siderite deposits of historical importance were located in the Erzberg (Ore Mountain) section of Austria (Noricum) and in the Weald (Sussex) and Weardale sections of England. Manganese is also a constituent of bog iron. See German steel, Mushet, Robert F., Weald of Sussex, Noricum, steel, and alloy steel.

Manganese oxide: Important replacement for iron oxide during the fining (decarburization) of cast iron into steel, as in the puddling furnace. Manganese oxide has a lower melting temperature than iron oxide as a slag constituent and facilitates uniform carbon distribution during steel production (Barraclough, 1984b, 91-106). See decarburization, manganese, and puddling furnace.

Manganese steel: In its modern form, an alloy steel containing 11 to 14% manganese and 1.0% carbon. It is used for safes, steam shovels, and other applications requiring high wear resistance. It cannot be machined or softened by annealing but is made somewhat softer and less brittle by quenching, the very reverse of the usual steel treatment process. See alloy steel, spathic ore, and spiegeleisen.

Manganiferous iron ores: Iron ores containing manganese.

Maraging steel: Modern *carbon-free* steel containing an iron-nickel martensite structure, which becomes hard and tough with aging and/or heat treatment after initially being ductile, malleable, and, thus, machinable. The high nickel content of maraging steels (18 – 25%) results in exceptional hardness and high resistance to stress corrosion (Pollack 1977).

Marriage of psychiatry with pharmaceuticals: Can ironmongers avoid ADHD, PTSD, depression, anxiety, psychoses by wielding their instruments of manual operation? Assault rifles anyone? See for-profit mental illness.

Martemper: To quench in molten salt at temperature slightly above the critical temperature to produce a fully martensitic structure in a uniform nearly stress-free manner.

Martenize: To form a hard, brittle, allotropic form of steel by the rapid cooling (quenching) of austenite. Martensite formation begins at a temperature of 220° C and below and contrasts with the formation of ferrite-pearlite microstructures from the slow cooling of austenite, which begins at 1000° C, or the wide variety of microstructures (bainite) formed by halting cooling and holding the metal alloy being cooled at temperatures between 1000° C and 220° C for various time periods. See allotropic forms of iron and austenite.

Martensite: "Martensite is considered to be a super saturated solid solution of carbon in ferrite; it is the hardest, strongest, and least ductile form of steel" (Shrager 1949, 141). Hard, brittle steel with various distorted microstructures determined by the range of carbon content and distribution, martensite is formed by rapid displacive transformation, rather than slower diffusive transformations. The microstructures are lath-like in low carbon steels and plate-like in high carbon steels. Derived from the rapid cooling (quenching) of austenite, tempering will alleviate brittleness and restore ductility and, therefore, usefulness in edge tool production. Martensite forms at lower temperatures than pearlite, which is prevented from forming by the rapid cooling of austenite. It has an unstable body-centered tetragonal structure differing from austenite and ferrite only in its axial ratio, which is determined by carbon content. Martensite is formed rapidly (0.002 seconds) when austenite is cooled below 240° C and has a needle-like crystallographic structure, with no diffusion into particles of cementite. During the tempering of martensite, carbon is diffused into finely dispersed particles of cementite by a decrease in its crystal structure axial ratio. Martensite can be considered an intermediate step in the transformation of austenite to ferrite (Barrett 1943, 478). See austenite, cementite, martensite, microstructure, quench, and temper.

Martensite shearing: In contrast to the diffusion of carbon during the tempering of martensite to produce malleable, ductile steel, martensite shearing is the result of a combination of the presence of alloys such as cobalt, vanadium, manganese, silicon, chromium, and tungsten and heat treatment processes. These processes result in formation of tangential stresses that change the angle between the face and the length of the diagonal of the microstructure of the alloy steel being heat treated, creating stressed (super hard) microstructures. Martensite shearing is characteristic of the phase transformation processes (kinetics), which result in the production of modern hardened steels. See alloy, microstructure, stress, and temper.

Martensite tempering: At temperatures between 220 and 450° C, some of the carbon leaves the martensite crystal structure and precipitates as carbides, leaving behind pure alpha iron ferrite, which provides more ductility as a component of martensite-ferrite-bainite microstructures. The tempering time to produce the complex microstructures that include bainite varies from ½ to 8 hours. After longer tempering times, pearlite structures begin appearing and dominate the microstructure complex (Pollack 1977). See the phase transformation diagrams in *Appendix VI*.

Martensitic transformation: "The coherent formation of one phase from another without change in composition by a diffusionless, homogenous lattice shear" (Petty 1970, 5).

Mass spectrometry: "…an analytical technique used to measure the mass-to-charge ratio of ions. It is most generally used to find the composition of a physical sample by generating a mass spectrum representing the masses of sample components. The mass spectrum is measured by a mass spectrometer" (Wikipedia 2008). Mass spectrometry is the characteristic measurement method used in all forms of electron microscopes. See electron microscope, TEM, REM, SEM, and HRTEM.

Matchlock: Device used to ignite a firearm; term used to describe some early firearms. The earliest form of a handgun was the arquebus using a simple serpentine matchlock ignition system. Some early muskets also used matchlock ignition systems. Matchlocks were unwieldy and unreliable and required a source of fire, such as a lit match, which would reveal the position of the gunner to his adversaries prior to the firing of the weapon. Nearly useless during rainstorms and inclement weather, the matchlock was always made of folded and forged strips of wrought iron. The battle of Saco, Maine, 1607, is a landmark event in American history, during which Micmac Indians, supplied by French traders, made first use of firearms in the form of matchlocks in a pitched battle defeating the Abenaki community during the fur trade wars. See the Davistown Museum publication *Norumbega Reconsidered* Appendices D and E (Brack 2006), arquebus, flintlock, wheellock, gun barrel iron, and gunsmithing.

Maximum permissible forging temperature (MPFT): Ranging from 1300° C for low carbon wrought iron, the MPFT declines with increasing carbon content, reaching approximately 1100° C for steel containing 1.2% carbon. Steel containing more than 1.5% carbon is "virtually unforgeable" (Barraclough 1984a, 5). See forging, incandescent temperature colors, steeling, and quenching threshold.

Mechanical properties of metals: Properties that determine the ability of a metal to withstand force without tearing, bending, or fracturing. See compressive strength, tensile strength, toughness, brittleness, malleability, ductility, impact strength, fatigue, elastic limit, and creep.

Mechanical working: The shaping of metal by hammering, rolling, pressing, punching, bending, or drawing to change its form. These processes may also change the metal's physical properties, i.e. its crystallographic structure. The fabrication of armor by a combination of cold hammering and heat treatment is the classic example of medieval era thermo-mechanical working of iron and steel. See lattice structure, cold forging, and cold working.

Melting furnace: Furnace used to re-melt smelted iron prior to its specific-purpose use. Fineries, chaferies, and cupola and puddling furnaces are all examples of early modern melting furnaces common before the advent of bulk process steel technology.

Melting points of iron and steel: See temperature range for melting iron and steel.

Merchant bar: High quality wrought iron bar stock re-processed in fineries by piling, reheating, and re-rolling bloomery-derived bar stock or puddling-furnace-derived muck bar. See ancony, bar stock, bloomery, chafery, finery, and muck bar.

Mercury: Chemical element in the periodic table with the atomic number 80 and the symbol Hg. Mercury is one of five elements that are liquid at or near room temperature. Mercury is among the most toxic of all forms of biologically significant chemical fallout (BSCF). With respect to ferrous metallurgy, its importance lies in the large amount of mercury released through the burning of coal, the smelting of iron, and the production of steel. Harmless in its naturally occurring form (mercuric sulfide), it undergoes transformation through bacterial action into highly toxic methylmercury, which is readily absorbed by living organisms and is easily transported in pathways to human consumption. Increasing rates of the cross placental transfer of methylmercury from mother to child has been documented in the late 20th century. Widespread contamination of fresh water and food products, including some farm-raised fish, may play a much more important role in the rising rates of autism in children than methylmercury contamination from specific source points, such as thimerosal, the antiseptic/antifungal once commonly used in children's vaccinations. Mercury is one of many skeletons in the closet of the pyrotechnic society. See BSCF, pyrotechnology, biocatastrophe, and infrastructure collapse.

Merovingian swords: Swords produced by Celtic metallurgists for the Frankish kingdoms preceding and during the reign of Charlemagne. Merovingian pattern-welded swords were famed for the quality of their steel and their artistic beauty. Predecessors of the later Viking all steel swords, Merovingian swords represent the high point of Celtic metallurgy, which utilized the natural steel produced in Noricum and transported down the Iron Road and then upstream on the Danube River to a portage west of Nuremberg. This portage linked the smelting furnaces of ancient Noricum to the metal smithing centers of the lower Rhine, supplying the famed swordsmiths of the Merovingian era.

Metal cutting technologies: Technologies for cutting metals developed on a parallel track with welding methods, often using the same tools. Early cutting was done with chisels, often on material hot from the forge, using a hot chisel or hot set, the sledge hammer version of a hot chisel. Cold cutting with a cold chisel or cold set had endless applications, often in tasks involving light gauge materials, such as in the manufacture of armor. Armorers would have made good use of oxy-acetylene torches for both welding and cutting, but they were not available until toward the end of the 19th century. So called cutting torches and their heavier relative the oxygen lance use the principle of rapid oxidation of ferrous material through the cut line, quite different than the removal of chips in the path of a tool, such as in a lathe-based system or even with a hacksaw. Handheld hacksaws were known to be used in medieval times, eventually leading to horizontal and vertical band saws that use endless blades usually of bimetallic construction. Tungsten carbide-tipped circular saw blades produce accurate cuts when mounted on stationary and portable machines. These are distantly related to abrasive cut-off systems that use circular blades, which grind through material and are consumed in the process. Other contemporary strategies involve an electric arc, as in the plasma arc machines now common in many shops, and some do not, as in the abrasive waterjet system, which will cut almost any material in addition to all metals (McLaughlin, David 2008, Liberty Salvage Co., personal communication).

Metallofullerene: A customizable metal-matrix composite consisting of self-replicating homochiral chains that allow for superior flexibility and resilience, high strength and durability, and an extremely low coefficient of friction. It also may be used in the realm of microRNA, the next step in pharmaceuticals (http://www.nascentnanotech.com/tech.html).

Metallographic analyses: Analysis of metals to determine their microstructure, including wrought iron, steel, and cast iron. The five principal techniques of metallographic analysis are: etching of metal surfaces with potassium bi-chromate, observation and photography of crystal structures by the optical microscope, spectrographic analysis, x-ray diffraction analysis, and x-ray fluorescence analysis. See photomicrograph.

Metallography: The study of the structure of metals and alloys and the properties that derive from their thermal and mechanical processing. As a science, metallography lagged several centuries behind the technological development of early modern steelmaking techniques (blister steel, 1685; crucible steel, 1742). Henry Clifton Sorby's pioneering microscopic analysis of the structure of metals did not receive widespread attention until 1882. See blister steel, crucible steel, microscope, microstructure, and Sorby, Henry Clifton.

Metalloids: Non-ferrous constituents of cast iron and steel up to 25% in volume, e.g. carbon in cast iron in the range of 2.2% – 5.0%; also silicon, sulfur, phosphorous, manganese, etc. Metalloid alloys in cast iron and tool steel, such as manganese, chromium, nickel, and other metals, radically change the microstructure and, thus, the appropriate function of the cast iron or tool steel being produced. See alloy, alloy steels and alloy cast irons, cast iron, Chalibean steel, manganese, and silicon.

Metallurgy: "The art and science of extracting metals from their ores and other metal-bearing products and adapting these metals for human utilization" (Shrager 1949, 371).

Metalworker: A modern generic term for ironmongers of all types: blacksmith, bladesmith, gunsmith, founder, etc.

Metaphor: A figure of speech. Microstructural deformation, stress field, slip interference, fracture, strain hardening, creep embrittlement, internal stress, interrupted quenches, lattice bindings, martensite shearing, overheating, and grain displacement are all metaphors for a global consumer society in crisis.

Metastable: Term describing any process that is characterized by rapid transformation, as in the rapid transformation of austenite to martensite (nearly instantaneous, but within a critical cooling rate of 80 seconds or less from 1000° C to +/- 200° C).

Methylmercury: Product of bacterial action on elemental mercury in the form of mercuric sulfide after its release by coal and coke-fired blast and steel furnaces and electric power plants. Methylmercury is one of the most potent and biologically significant ecotoxins in the repertoire of the global contaminant pulse. It is the skeleton in the closet of a pyrotechnic society that uses biomass to fuel its heat engines. See mercury, biocatastrophe, infrastructure collapse, and post-apocalypse society.

Micrographs: Data produced by optical and electron microscopes during metallurgical analysis, initially in the form of photomicrographs and, after the invention of electron microscopes, by x-ray diffraction measurements of the microstructure of metals. See electron microscope, TEM, HRTEM, REM, and SEM.

Microscope: A tool utilizing one or more glass lenses to view objects too small to be seen by the naked eye. The invention of the microscope was the first technological innovation allowing analysis of the crystal structure of metals. First developed in the Netherlands in the early 17th century, microscopes were initially used to investigate the structure of biological materials. Henry Clifton Sorby (b. 1826, d. 1908) inaugurated the science of metallography with his etching and microscopic analysis of sectional slices of various ferrous metals (wrought iron, cast iron, Bessemer steel, blister steel, decarburized cast iron, spiegeleisen) (Smith 1960, 173). See electron microscope, HRTEM, optical microscope, and Sorby, Henry Clifton.

Microstructures: In ferrous metallurgy, an almost infinite variety of the morphologies of the crystal structures in the various phases of the carbon iron system, resulting from the thermal and mechanical treatment of steel and iron. Microstructures can be observed by optical and electron microscopes. Microstructure determines the forms and properties and, ultimately, the uses of iron and steel. See annealing, austenite, crystal structure, electron microscope, lattice structure, optical microscope, martensite, and unit cell.

Microstructure variables: The microstructure of various allotropic forms of steel is contingent upon an almost unlimited number of variables with respect to thermal and mechanical treatments. Generic variables include temperature, cooling rate, carbon content, and alloy content. See bainite, crystal structure, googolplex of microstructure, mechanical working, and thermal treatment.

Mild steel: Another name for bulk-process-produced low carbon steel (0.1% - 0.3% cc). Before the era of bulk process steel (1870), malleable iron, characterized by the same carbon content as mild or "low carbon steel" but including siliceous slag as a micro-constituent, served many of the same functions (e.g. toolmaking) as "modern" mild steel. See carbon content of iron and steel and low carbon steel.

Mill forge: To create directional grain structure by the action of rollers under great pressure in modern rolling mills.

Millscale: The oxidizing agent, iron oxide. Millscale is initially formed from the hammering and rolling of heated iron, the surface of which is oxidized to form magnetic iron oxide. Millscale, along with lime and fluorspar, is often used as a flux in smelting, and, as an oxidizing agent, greatly increases the efficiency of puddling furnaces. See flux, oxidation, and reduction.

Minimum quenching temperature: Temperature below which rapid quenching will not result in hardening; also called the critical point. See austenite, martensite, and quenching.

Miyairi: Japanese swordsmith.

Modern era: With respect to ferrous metallurgy, that period of time characterized by the perfection of bulk process steel production (>1875); also that period of time beginning after Benjamin Huntsman reinvented a method for producing crucible cast steel (1742). With respect to the Industrial Revolution, the modern era is also denoted by the widespread appearance of Watts' steam engine (1775), which involved the innovation of doing work by boiling water rather than harnessing the power of falling water. The modern era is also denoted as the Age of the Factory System of Mass Production (>1840), and lastly, as that historical era characterized by the development of the electrical transmission grid and the automobile (>1890), i.e. the Era of Petro-Chemical Electrical Man. See cascading Industrial Revolutions.

Modes of carbide dispersion: Upon the cooling of austenite, iron carbide may form "in plates (lamellae), intergranular cells, or somewhat spherical particles, all over a vast range of sizes. Nowhere in metallurgy is there a more complete representation of the effects of the distribution of hard particles in a soft plastic matrix than in steel. The finer the dispersion, the greater the hardness of the aggregate" (Bain 1939, 22).

Mohammed's ladder: Special pattern of repeated vertical markings on Damascus steel swords. These markings are a result of the distribution of ultra fine networks of iron carbide during warm forging of the Wootz steel in Damascus swords. See Damascus steel, nanowires, nanotubes, ultra fine microstructures, superplasticity, and Wootz steel.

Moiré Métallique: Intentional use of patterns resulting from crystal formation to produce decorative effects on metals. Invented by Allard in 1814, etched tinned iron plate was covered with a lacquer to produce lacquered tinware (Smith 1960).

Mokumé: Disclosure of the structural and chemical properties of pickled iron and steel in the form of wood grain patterns for the purpose of decorating Japanese iron sword guards (tsuba). Described and illustrated by Cyril Smith in his chapter on Japanese swords (Smith 1960, 57-62).

Mold: Form that contains the cavity for the casting of metal objects from molten metals.

Molded malleable cast iron: Malleable cast iron, which, after annealing, is suddenly chilled in iron molds, producing a steely cast iron with a very smooth surface, as in Bailey plane bodies c. 1855. Malleable cast iron may have been subject to additional heat treatments to become grey cast iron, which was used by the Stanley Tool Co. for planemaking. An exact description of the heat treatment process used to manufacture high quality Stanley iron bodied planes in the last four decades of the 19th century remains elusive.

Molder: In the operation of the blast furnace, the person in charge of making sand molds out of wooden patterns, which formed the shape of the cast iron object being manufactured, and then supervising the pouring of the melted cast iron into the mold. See blast furnace, hollowware, and patternmaker.

Molder's finishing tools: Small steel hand tools used for shaping, smoothing, and finishing the surfaces of sand molds, e.g. slicks, corner slicks, trowels, and lifters.

Molybdenum: Most effective of all alloys in the hardening of steel by phase transformation kinetics (Cias n.d.).

Mountain ore: Term used by Peter Oliver to refer to iron ore being imported, preferably from Pennsylvania, to be used as a substitute for the rapidly disappearing bog iron deposits of the Middleboro, MA, area. Oliver's Forge on the Nemasket River in Middleboro was then producing mortars, cannon, and howitzers for the Crown. In a letter dated March 21, 1756, Oliver, a Tory, noted the superiority of "mountain ore" over bog ore for heavy ordnance (Weston 1906).

Moxon: Joseph Moxon's *Mechanick Exercises or the Doctrine of Handy-Works* [1703 (1989)] is the most comprehensive survey of blacksmithing, toolmaking, and crafts such as joinery, house carpentry, turning, bricklaying, and mechanical drawing available at the end of the 17th century. It contains important illustrations of basic tools, some of which, such as the frame saw, bow saw, scribe, bevel, keyhole saw, chisels, and gouges, are still found in old tool chests and collections, linking contemporary tool forms to English prototypes used in the early North American colonial period.

Muck bar: Nineteenth century American term for puddling-furnace-derived, unprocessed wrought iron produced by a single passage through a rolling machine. Muck bar was usually subject to further refinement by additional shingling, rolling, or faggoting and piling into higher quality wrought iron. Also used in this Glossary in reference to the bar stock produced by the bloomsmith prior to refining.

Multiple alloys: Alloys that play a role in phase transition kinetics by enhancing or mitigating mechanical properties such as hardness.

Multiple hearths: While North American colonial era forges and smelting furnaces often had only one hearth, most integrated ironworks had multiple hearths, occasionally as many as twenty or thirty. Ancony hearths, chaferies, blacksmith forges, and anchor forges are examples of special use hearths used at integrated ironworks or colonial forge complexes where the multiplicity of hearths differentiated these ironworks from the community forges of individual shipsmiths, edge toolmakers, and blacksmiths.

Multiple transformation modes: Modes occurring within the critical cooling rate temperature range (1000° F to 200° F) and resulting in the production of a wide variety of morphologies, ranging from lamellar structures to structures that exhibit spheroidization at grain boundaries. Alloying elements and the grain coarseness of the austenite being cooled also influence the multiplicity of microstructures formed within the critical cooling rate range. Bainite is the generic name for many of the microstructures formed within this temperature range and is a testament to Edgar Bain (1939), who first elucidated these phenomena.

Mumbra: Ancient African tribe that made cast iron using horn and quartz as flux (Wertime, 1962).

Mushet, Robert F.: (1812 - 1891). English metallurgist with a clandestine steelworks in the Forest of Dean. Mushet was the first English ironmonger to commercially produce spiegeleisen, an iron-carbon-manganese alloy, which he convinced Henry Bessemer to use in his conversion process for the mass production of low carbon steel. Mushet later played a major role in the development of alloy steels, inventing what he called self-hardening steel (1868). Initially made with high tungsten content, as well as manganese and other alloys, "self-hard" was the principal form of tool steel used for the next three decades. The metallurgists Taylor and White eventually discovered that manganese, rather than tungsten, was the principal alloy providing hardness. See manganese, R.M.S., self-hard, spiegeleisen, Taylor-White, and tempered alloy steel.

Music wire: Fine steel wire used as strings in musical instruments, as well as for precision measurement functions; produced by drawing annealed steel wire through a draw plate. See draw plate.

Myth of wrought iron: Mistaken belief that "until the iron makers of Central Europe developed a new type of furnace around 1350, wrought iron had been the only product made from iron ore" (Aston 1939). This is incorrect. The wide variety of iron tools, weapons, and currency bars recovered from early Iron Age Roman era and migration period archaeology sites indicates that wrought iron production (0.02 - 0.04% carbon content) in direct process bloomeries was rarely achieved. The great majority of artifacts recovered from these sites contain widely varying carbon contents ranging from what today would be called low carbon steel (> 0.08% cc) to heterogeneous raw steel with a carbon content as high as 1.5%.

Naginata: Form of Japanese weapon; a wood shaft with a curved blade and a tsuba (guard) between the blade and the shaft. Traditionally used by samurai warriors, the naginata is the subject of the modern ritual of naginatajutsu, especially popular with contemporary women in Japan.

Nanotechnology: Study of the microstructure of metals, ceramics, and other materials utilizing the high resolution transmission electron microscope (HRTEM), which can provide images of formations in the order of magnitude of billionths of a meter. Nanotechnology is especially useful in the study of the crystal structure of forged and heat-treated steels and the measurement of trace amounts of ecotoxins in pathways to human consumption. It is also associated with semiconductor physics of micro-fabrication technologies and potentially significant environmental contamination, which is the hidden downside of innovative nanotechnology. See nanotubes, nanowires, and picotechnology.

Nanotubes: With respect to ferrous metallurgy, the spheroid carbide (cementite) microstructures containing nanowires created by forging high carbon Wootz and other steels at temperatures below the eutectoid point. The carbide microstructure of nanotubes, which can only be examined using the high resolution transmission electron microscope (HRTEM), differs from the microstructure of pearlite, which has a lamellar morphology composed of alternating plates of ferrite and cementite. In ferrous metals characterized by the presence of nanotubes, the nanotubes, containing cementite nanowires, are often adjacent to plates of pearlite and provide elements of superplasticity not present in steels characterized by other microstructures. See carbon nanotube, nanowires, superplasticity, and Wootz steel.

Nanowires: Characteristic linear arrangement of carbide (cementite) structures within the nanotubes of superplastic steels, often in the form of parallel wires in areas *outside of* cementite particles. The distribution of nanowires is not uniform and their "preferred direction" can have local variations. "…there are obvious similarities between the nanowires found here and the laminar cementite in pearlite" (Levin 2005, 913). Nanowires may also contain other alloys, such as vanadium, cobalt, chromium or manganese, as micro-constituents, which may play a role in facilitating superplasticity. See nanotubes, spheroidization, and superplasticity.

Natural steel: "Steel made in a bloomery directly from iron ore. It contained slag particles since it was a carburized form of wrought iron" (Gordon 1996, 310). Natural steel production was dependent on higher than normal fuel to ore ratio, rapid air flow to facilitate high furnace temperatures, high silica, high manganese, and low phosphorous ore content (Barraclough 1984a, 17) and was the most common steel-producing strategy before the appearance of the blast furnace and the consequent development of the decarburization of cast iron to produce (German) steel. See bloomery, direct process, ferrosilicate, German steel, and manganese.

Natural steel edge tools: Tools that are forge welded (hammered, quenched, and tempered) by the blacksmith directly from the bloom of unrefined smelted iron. Starting with malleable iron bar stock with at least a 0.2% carbon content, additional carburization of the cutting edge of a tool results from its careful placement with a charcoal fire. Knowledgeable blacksmiths could make high quality edge tools using this method but only one or a few tools at a time. The best natural steel tools were made from manganese-bearing spathic ores, first at Halstadt (Austria), then by the Romans in Austria (Noricum), in Spain in the medieval period, and especially in southern Germany after 1400 AD. Natural steel edge tools were also produced, but probably with greater difficulty, by American blacksmiths utilizing manganese laced, bog iron-derived bar stock in situations where German or blister steel bar stock was not available for "steeling" edge tools. Natural steel was the most common constituent of forge welded edge tools from the early Iron Age until the late medieval period. See forging, bloomery, direct process, natural steel, and nest carburization.

Nest carburization: Steeling of groups of malleable iron chisels, caulking irons, or other tools by their submergence in charcoal pits for periods of days (24 – 72 hours). In this ancient variation of case hardening, nests of tools would often be enclosed by clay or other non-combustible coverings, which would protect the iron components of tools being steeled from deleterious effects of combustion gases. See carburize, case hardening, and enclosure.

Neumann bands: Bands observed in the microstructure of steel that result from mechanical treatments, such as hammering at room temperature or slow deformation at lower temperatures. These bands are formed by twinning, i.e. the grouping of identical lamellae in patterns characteristic of the crystal structure of a specific allotropic form of an iron carbon alloy, e.g. pearlite. See Widmanstätten structure, deformation history, plastic deformation, lamellae, lattice structure, bainite, and pearlite.

Newcastle: Principal steel-producing center of England between 1700 and 1775. Located in the far northeast of England, Newcastle predated Sheffield's emergence as England's late 18th and 19th century center of steel production. Newcastle emerged as an important steel-producing center after iron and steel production activities ended in the Forest of Dean and the Weald of Sussex. See Forest of Dean, Weald of Sussex, and Weardale.

Nickel: Important alloy in steel. Nickel helps retain austenite in high carbon steels except in those with a high chromium content (Pollack 1977). See Chalibean steel, chloanthite, and nickel steel.

Nickel-Silver: 50% copper, 25% zinc, 25% nickel; also called German silver; strong, hard, and corrosion-resistant.

Nickel steel: Alloy steel containing nickel; nonmagnetic with high tensile strength. Nickel "increases strength by strengthening ferrite without decreasing ductility" (Shrager 1949, 187). A typical nickel alloy steel is 3.5% nickel and 0.5% carbon and has several thousand pounds per square inch more strength than pure steel. Steels with more than 22% nickel are nearly corrosion resistant; those with 25% to 46% nickel are used specifically for watch parts and measuring instruments due to their failure to expand and contract. Meteor iron often contains about 10% nickel, which greatly strengthens it. Chalybean nickel steel has historically been misinterpreted as meteor-derived (Piaskowski 1982). See Chalybean steel and German steel.

Niello: Black alloy of silver, lead, and sulfur used for filling in engraved designs in silver and other metals. See alloy and chasing.

Nitriding: The modern process for surface (case) hardening selected alloy steels by heating the steel in an atmosphere of nitrogen (ammonia gas) at approximately 950 °F, followed by slow cooling. Heat treatments, including annealing, are usually applied before nit riding, which reduces distortion. Standard alloy steels containing chromium and molybdenum and nonstandard alloy steels containing aluminum are the only steels suitable for nitride formation. Iron nitriding is a recently developed process used to case harden cutting tools utilizing ionized low pressure gas.

Nodular cast iron: Made from iron with a low sulfur content, with various alloys such as sodium, magnesium, and cerium, in which carbon is present in spherical nodules, also called ductile cast iron. A versatile modern form of cast iron available in a variety of chemical configurations, characterized by high tensile strength and a significant capacity for elongation, nodular cast iron is representative of the wide variety of iron carbon alloy combinations developed in the 20th century. See cast iron.

Noric steel: Steel produced by Celtic metallurgists from the siderite ores of the Mount Erzberg region of Austria and supplied to both the Roman Republic and the Roman Empire for weapons production. Noricum was a Roman province, hence the name of the natural steel produced there from the manganese-rich spathic ores. See natural steel, German steel, Erzberg, Styrian steel, Celtic metallurgy, iron road, and spathas.

Noricum: Area of Austria called Styria and Carinthia including Mount Erzberg (Ore Mountain) that was the Celtic center of Roman ferrous metallurgy.

Normalizing: A heat treatment for steel. Normalizing is the slow cooling of iron-based alloys from temperatures above the critical point in an environment of still air and normal temperatures.

Norwegian steel: Forged and laminated steel made out of high carbon hard tempered steel and soft iron, having a Rockwell hardness of 63 – 64; used by knife makers and capable of being bent in a vise (Latham 1973).

Nucleation: "The formation, through thermally activated fluctuation, of the smallest stable particles of a new phase" (Sinha 2003, 6.1). Nucleation is characterized by randomness, rapidity of transformation, and is essentially equivalent to freezing. See freezing point of iron.

Nucleation capacity: Actual controlling factor in hardenability for a given composition; correlated with the austenitic grain diameter and the presence of alloying elements, which change the composition of austenite prior to hardening (Bain 1939, 54). See phase transformation and kinetics.

Oakum: Shredded rope or hemp fiber often used for ship caulking, thus requiring the use of caulking irons and mallets, standard tools of the working shipyard. See caulking iron.

Oil temper: In the forging of edge tools, temper obtained by the quenching of tool steel in oil rather than brine or water. See hardness scale, Rockwell hardness test, temper, martensite, and brine quenching.

Open-hearth furnace, early forms: The most common form of furnace used in antiquity, the open-hearth furnace had two basic forms, bowl and shaft, both of which exposed the ore being smelted to the direct impact of the oxidizing flame of the burning fuel. The shaft furnace, which dates to the early Iron Age in central Europe, was later developed into the high shaft German Stucköfen furnace, the predecessor of the blast furnace. The bowl furnace, which later evolved into the Catalan furnace, was used in southern Europe until the early 20th century. Both were used for direct process production of iron blooms, and both forms occur as slag pit furnaces and slag-tapped furnaces. The most primitive slag pit type furnaces were characterized by the accumulation of slag in pits underlying the furnace, which required the movement of the furnace to adjacent slag-free locations. More practical and productive were furnaces designed with adjacent pits for slag tapping, the residue of which was more easily moved to rubble piles in the form of slag heaps. See furnace forms, shaft furnace, bowl furnace, slag pit furnace, slag-tapped furnace, bloomery, and direct process.

Open-hearth furnace, modern forms: Another name for two modern (after 1865) forms of the blast furnace, i.e. the Siemens-Martin process and the Siemens open-hearth furnace, which are differentiated primarily by the type of material used in the charge. The Siemens-Martin process used pig iron and scrap steel for the charge; the Siemens open-hearth furnace used carefully measured additions of iron ore to the slag after the melt was underway, increasing its iron oxide content. Both use waste gases to preheat the air blast and are considered superior to the Bessemer blast furnace because they allow better quality control of alloy steel production. The open-hearth furnace process is also less wasteful than the Bessemer furnace, utilizing all ores and scrap, which were more slowly heated to higher temperatures by indirect radiated heat from a regenerative gas furnace. Modern open-hearth furnaces use gas, coke oven gas, natural gas, oil, and powdered coals as possible fuels. In the Siemens open-hearth process, "molten pig iron is run onto a hearth or bedcover, which gas flames play to maintain the heat and to burn up carbon and other defects [contaminants]. This process took some eight hours, and in the result, was able to produce an even nature throughout" (Abell 1948, 148). See basic process, acid process, Siemens-Martin process, Siemens open-hearth furnace, and the Siemens regenerative gas furnace.

Optical microscope: Microscope used by William Sorby (c. 1885) to investigate the space lattice structure of ferrous metals. The optical microscope inaugurated the science of metallography, i.e. the study of the microstructure of metals.

Ordnance: Military supplies and weapons. Bronze and cast iron cannons and artillery were the most important product of foundries and ironworks producing ordnance in 16th century England and Europe. For example, the casting of cannons for the coastal defenses of England was the most important use for the products of the blast furnaces of the Weald. The demand for ordnance, gun iron, shipsmith's ironware, and edge tools resulted in the expansion of the English iron industry from the Forest of Dean and the Weald north to the midlands after 1600, to Newcastle after 1700, and to Sheffield after 1750. Before 1635, the location of production of the finest early handguns by gunsmiths and sword cutlers was southern Germany (Nuremburg, etc).

Ore boil: The violent frothing that occurs in the open-hearth process as a result of the interaction between the iron ore and iron oxides in the slag and the carbon in pig iron. The carbon monoxide thus formed causes the frothing. In the puddling furnace, the termination of the boil signals the full decarburization of the pig iron and its conversion to wrought iron. See puddling furnace.

Oriental crucible steel: Produced in high temperature resistant refractory clay crucibles in small quantities (+/- 3 kilograms) in India, China, and Japan as early as 300 BC and possibly earlier in India. Crucible steel was a trade item which may have reached the Levant in the first millennium and was probably used by the Vikings in sword production. Damascus steel is one variation of Oriental crucible steel. When Benjamin Huntsman "discovered" crucible steel, he was readapting an already known, but obscure and difficult, steelmaking process for his watch spring manufacturing business. See cast steel, Wootz steel, and Venetian steel.

Osmund: A form of malleable iron or natural steel imported from Sweden and Spain during the Medieval period and used for knitting chain mail, also used to make arrowheads and fishhooks.

Over-coating: A modern method of ax-making; the overlaying of a piece of high carbon steel (the ax bit) upon a folded iron or low carbon steel ax body prior to welding (Kauffman 1972, Klenman 1998). See ax-making techniques and iron cored.

Overheating: Process that results in the surface decarburization of iron or steel, which begins occurring at a temperature of 1200° C. Surface carbon is removed and the iron is oxidized into an iron oxide scale. "No useful degree of carburization can be achieved when iron is heated unprotected in a forge fire" (Barraclough 1984a, 31). Overheating in direct process smelting had the ironic effect of promoting carbon uptake in the bloom, facilitating the formation of liquid cast iron. Iron absorbs carbon more rapidly at high temperatures; the increase in the carbon content of the bloom then lowers its melting temperature. See enclosure, bloomery process, natural steel, cast iron, and carburization.

Overseas carrying trade: A term referring to the North American colonial era transport of masts, spars, planks, and other timber products to English and continental shipyards.

Overshot water wheel: The most efficient form of the water wheel, which utilizes the height and volume of falling water as the power source for the operation of the blacksmith's trip hammer, the up and down mill saw, the grist mill, and belt-driven machinery of every description. See undershot water wheel.

Oxidation: The combination of oxygen with an element to form an oxide or the combination of an oxide with more oxygen to form a higher oxide (Shrager 1949). During the smelting process, oxidation removes impurities from the higher iron oxides, resulting in the production of a lower form of iron oxide, FeO, the precursor to the formation of pure wrought iron (ferrite, Fe).

Oxidation reduction reaction: The simultaneous occurrence of the oxidation and reduction of iron in the smelting process. Intermediate reducing agents remove oxygen from the reductant, e.g. iron oxide, therefore oxidizing it. Carbon removes oxygen from iron oxide and then is transformed into carbon monoxide: $FeO + C = Fe + CO$. Both processes work ideally to produce the ferrite (Fe) sought by the bloomsmith. Entrained in this ferrite is siliceous slag (ferrosilicate) and the combination of the two constitutes wrought iron. With the exception of sulfur, oxidation results in the removal of most forms of impurities as slag. See ferrosilicate, oxidation, reduction, and smelt.

Oxidizing reaction: In iron-smelting, the change in composition that occurs during the making of wrought or malleable iron or steel from iron ore or pig iron. The carbon in the fuel separates as a gas in the form of CO or CO_2 and thus removes oxygen from iron oxide, producing iron or steel. When ferromanganese and ferrosilicon are added to the steelmaking charge, they serve to reduce the dissolved gases and the iron oxide in the metal. The manganese and silicon form oxides, which collect as slag and encourage the uniform distribution of carbon in iron to produce raw steel. See natural steel.

Oxygenation: The reaction of oxygen with hot metal (e.g. iron oxide) to produce CO and CO_2.

Pack carburization: 1) The case hardening of stacked tools packed in a metal box containing carboniferous material or in a pit containing charcoal. 2) The sealing of iron or steel in a furnace or container in conjunction with carboniferous materials; an ancient strategy for carburizing iron or low carbon steel that is still used in the modern era. The release of carbon diffuses below the surface of the iron or steel to form carbides when the steel is heated above the critical temperature. See case hardening, cementation steel, and nest carburization.

Packing: The compacting of fibers of wrought iron or carbon steel by hammering in order to increase the density and, thus, the strength of the tool or implement being forged. For example, packing is done with carbon steel when the metal is "sunrise red, barely discernible in regular light… if the metal is too hot, it will not pack, if too cold it may crystallize and break" (Bealer 1976a, 168).

Palimpsest: In a historical context, a historical palimpsest is a parchment written upon once and then covered over again by subsequent narrations that obscure the original inscriptions. In an archaeological context, the palimpsest is the accidental durable remnants (ADR) used to interpret prehistoric cultures and consists of the multiple layers of material cultural artifacts, the archaeologists' excavation and study of which results in the compilation of an account of these societies. The primary source of information about prehistoric societies are the forms of tools and other artifacts used and discarded in layers of cultural materials, now the subject of study by archaeometallurgists and archaeologists.

Pattern books: Toolmakers' catalogs issued in England by manufacturers, such as Wykes (watch making tools), Stubbs (files and hand vises), and Timmins (tools of the trades), which often illustrated tool forms derived from Roman and near eastern sources. Pattern books signal the rise of a market economy in the late 18th century when the gentleman and his tool chest become equal, if not greater, in importance to the lowly artisan, such as the file-maker, blacksmith, or shipsmith, all of whom used tools to make other tools. See Diderot's *Encyclopedia*, iconography of tools, and Moxon.

Patternmaker: The woodworker who crafts the wooden patterns that form the sand molds for casting bronze, brass, aluminum, cast iron, and steel into the wide variety of tools and equipment made in foundries. See cope, core box, and drag.

Patternmakers' shrink rules: Rules that, by graduated variations in their increments, make allowance for the changing (shrinking) dimensions of metals, which contract at various rates during the cooling process from the freezing temperature of the hot metal to room temperature. With as many as 12 variations of allowed shrinkage to the foot ranging from 1/12" to 3/8" (Stanley 1984), such shrink rules provided for the fabrication of larger patterns necessary to compensate for the tendency of metal to shrink upon cooling.

Pattern-welding: The ancient craft of knife and sword production based on welding together strips of natural steel or case-hardened steel and strips of wrought iron, which were forge welded, quenched, and tempered to produce the desired weapon; used as a sword and knife-making technique from the early Iron Age to the modern era; made famous by Merovingian, Viking, Muslim, and Japanese sword makers. See Damascene, reagent, sword cutler, and steel smith.

Pearlite: A mixed layer structure of iron carbide (cementite) and ferrite; a microstructural form of steel derived from the eutectoid transformation of austenite. Pearlite has a much higher tensile strength (125,000 lbs/sq inch) than ferrite (50,000 lbs/sq inch). If the lamellar morphology of alternating plates of ferrite and cementite containing 0.83% carbon is cooled below 723° C, it then becomes pearlite (Wayman 2000). "Pearlite is in fact the dominant microstructural form in medium to high carbon steels (0.4 – 0.8 percent C)" (Sherby 1995, 11). See cementite, ferrite, lamellae, and microstructure.

Pearlite nodules: Nodules that contain multiple colonies of parallel lamellae characterized by random arrangements, often as interwoven bicrystals of ferrite and cementite; a common characteristic of pearlitic steels.

Pearlitic malleable cast iron: A cast iron, the manganese content of which allows its pearlite matrix to retain carbon in the form of cementite.

Persistent organic pollutants (POP): Long-lived, synthetic, organic chemicals produced by pyrotechnic societies in large quantities beginning in the middle of the 20th century. Most biologically significant POPs are volatile and/or lipid soluble, bioaccumulate in pathways to human consumption, and are globally transported by hemispheric transport mechanisms, including the water and carbon cycles. See biocatastrophe and biologically significant chemical fallout.

Pewter: An alloy of tin and lead or tin and copper. Bismuth and zinc are also sometimes a constituent of pewter. While Roman pewter was high in lead, modern pewter, if intended to be used as an eating utensil, is made without lead. See alloy.

Phase: With respect to ferrous metallurgy, a physical (allotropic) form of iron characterized by specific microstructures. "Phase is a portion of a system whose properties, composition and crystal structure are homogenous and which is separated from the remainder by distinct boundary surfaces. …the phases present in Fe-Fe_3C and Fe-C diagrams are molten alloy, austenite, α-ferrite, δ-ferrite, cementite, and graphite" (Sinha 2003, 1.3-4). See allotropic forms of iron, *Appendix VI*, phase transformation kinetics, and phase transition.

Phase diagram: Diagram that expresses the relationship of carbon content to temperature; also called an equilibrium diagram. Temperature is plotted vertically and carbon content is plotted horizontally. See *Appendix VI* for a sampling of the wide variety of phase diagrams and an illustration of the simple phase transformation kinetics of pure iron-carbon formulations.

Phase transformation kinetics: The study of the process of altering the microstructure of ferrous metals through mechanical and thermal treatments and by the addition of alloys; especially used in modern steelmaking strategies to create hardened steels.

Phase transition: The change in the microstructure of iron from one allotropic form (e.g. austenite) to another allotropic form (e.g. pearlite) without a change in its chemical constitution, usually due to thermal and/or mechanical treatments.

Phenomenology of tools: The sum total of all the ways in which we construct and deconstruct our world with tools, i.e. the language of tools, including not only the making and using of tools but also the construction of landscapes and cityscapes as an iconography of tools at work.

> The phenomenology of tools is the totality of the voices of tools in history, a large extended family of function, skill, technique, necessity, history, obsolescence, memory, expressed in a cacophony of images and artifacts, all squabbling to be used or laid aside, remembered or forgotten. From eolith to atomic bomb, from artisans to nation states, the craftsmanship and history-making of tools join with the archaeology of tools to construct, deconstruct, and again reconstruct the stories of history. (Brack 2005, 7)

The phenomenology of tools is, in essence, the writing of history by the wielding of tools, as well as the inscription and incorporation of tools in and as art. In the 18th and 19th centuries, huge areas of North American forest were cleared by timber harvesting, charcoal production, and farming. Later in the 19th century, much of this landscape became towns and then cities with paved roads. The transition from handmade hand tools to machine-made hand tools to a global consumer society, where almost all artifacts are machine-made, characterizes the evolution of industrial society. Implicit in this dynamic of the phenomenology of tools are the lessons of pyrotechnic societies, i.e. the recurring narrations of the deleterious environmental impact of carbon dioxide emissions (Gore 2005), biologically significant chemical fallout, or the looming possibility of nuclear disaster, antibiotic resistant infection, biocatastrophe, and infrastructure collapse. What is it, in fact, that we are constructing or have constructed with our tools? See biocatastrophe, Brack (1984), infrastructure collapse, methylmercury, and post-apocalypse society.

Phlogiston theory: A theory popular between the late 17th and late 18th century that stated that combustion was due to the presence of phlogiston in the flammable materials, the phlogiston having no mass, taste, odor, or color and being produced as ash by combustion. It was assumed to occur in all flammable materials and to be released by the combustion process. The discovery of carbon and its role in combustion put an end to the phlogiston theory.

Phosphorus: Inflammable, nonmetallic, chemical element (P) commonly found as a naturally occurring contaminant in iron ore. When present in wrought iron (+/- 0.025% cc), phosphorus increases hardness, making it useful for nail-making. It increases the fluidity of cast iron and thus aids in the casting of hollowware. Otherwise, phosphorous is not useful as a hardener when in the presence of carbon; the higher the carbon content of malleable iron or steel, the more deleterious the phosphorous content. Phosphorous content above (+/- 0.05%) causes brittleness at room temperature. See cold short.

Photomicrograph: A photograph of a microscopic or magnified object taken with a microscope. The appearance of more powerful light microscopes in the late 19th and early 20th centuries allowed detailed analysis of the wide variety of the crystalline microstructures characteristic of various allotropic forms of ferrous metals. Such photomicrographs led to the discovery that pearlite was a form of steel composed of alternating lamellae of ferrite and iron carbide (cementite). See allotropic, cementite, electron microscope, ferrite, HRTEM, iron carbide, lamellae, microscope, optical microscope, and pearlite.

Pickling: Use of a diluted acid bath to remove oxide scale from iron and steel surfaces prior to mechanical treatments, such as cold rolling or wire drawing.

Picotechnology: The study of structures, including those of ferrous metals and ecotoxins, in parts per trillion (sub-nanometer quantities), where measurement of single atoms or molecules in metals and especially in abiotic and biotic media will provide additional information about the microstructures of metals and, most importantly, the movements of ecotoxins in pathways to human consumption. Picotechnology will be particularly useful in documenting the presence of persistent organic pollutants (POPs) in environmental media, which are currently below the limit of detection of most nanotechnological analysis systems at lower trophic levels. Many ecotoxins are biomagnified at higher trophic levels and bioaccumulate in living organisms, reaching their maximum level of concentration and synergistic interaction in wildlife feeding at the highest trophic level (e.g. bald eagles, humans).

Pig iron: The high carbon iron silicon alloy (2.0 - 5.0% carbon content) produced by the hot temperatures of large shaft furnaces and, later, blast furnaces. The hotter the furnace fire, the more rapid the absorption of carbon by the iron being smelted. Originally a waste product of smaller bloomery and shaft furnaces, pig iron is the first stage product of the indirect process of iron- and steelmaking. Because cast iron was produced in a liquid form in the blast furnace, when it was drained off it was solidified in rows of sows. Hence, the term "pig" iron was due to its appearance as cooled ingots. "Iron alloyed with carbon and silicon produced in a blast furnace" (Gordon 1996, 310).

Pile: To bundle and reforge iron or steel bar stock.

Piled wrought iron: A term for refined wrought iron made by the repeated faggoting of bars or plates of bloomery iron or puddled wrought iron, followed by shingling (hammering) to remove additional carbon and slag constituents and further align the silicon slag content into the fibrous patterns that provide its unique toughness and ductility. The high quality Swedish iron bar stock used in England and America to produce steel was highly refined, piled wrought iron, which was essential to the production of the highest quality edge tools. See finery, loup, refined wrought iron, shingling, and whalecrafters.

Piling: A technique used by knife- and swordmakers in all ages. Piling was the bundling and then forging of alternate layers of wrought iron and steel. In the age of blister steel (1686 – 1750) in England, piled blister steel was reforged into shear steel. The highest quality steel was known as double shear steel made by re-piling and reforging shear steel yet again (Barraclough 1984a). Piling also refers to the reheating of iron bar stock on clay tiles prior to reforging and the reheating of stacks of bar iron in a balling furnace. Shear steel production was gradually replaced by the crucible steel process, i.e. the production of steel by a chemical reaction that required fewer steps and less labor and fuel and offered more quality control of the slag and carbon content of the cast steel being produced. See knife forms, pattern-welding, and swordmaking.

Piling and folding: A process used by sword cutlers and edge toolmakers to reduce the heterogeneity of sheet steel and sheet iron during edge tool and sword production. Sheets of steel and iron were forge welded into a more homogenous mixture by piling and folding before reforging. The highest quality samurai and Damascus swords were piled and folded hundreds of times before a final product was ready for exchange. See Japanese sword, pattern-welding, and reagent.

Pin the boundaries: To form very fine grains of pure iron and iron carbide that prevent growth of large carbide spheroids, allowing the unhindered boundary sliding necessary for superplasticity in metals (Sherby 1995). See superplasticity.

Pins: Inclusions of unwanted natural steel in wrought iron not fully decarburized, making the iron useless for many applications. Pins were the bane of furnaces making iron shafts and blacksmiths making wrought iron hardware and implements. See natural steel.

Pipe: The shrinkage cavity within a cast steel ingot formed as a result of uneven cooling. In crucible steelmaking, this cavity is eliminated by the teamer, who has an extra reserve of liquid steel to put in a dozzle (clay cone) to fill the cavity. See cast steel, dozzle, and killed steel.

Piston bellows: Bellows developed in place of accordion bellows after 1750 and used in conjunction with steam engines. The higher efficiency of piston bellows allowed for the construction of larger blast furnaces. Community blacksmiths and shipsmiths continued to use accordion bellows to provide the air blast for their small forges until the late 19th century.

Pit vault molds: Molds set vertically in the ground to receive cast iron or bronze for cannon casting. Pit vault molds were Henry the Eighth's molder's principal method of casting cannon in the Weald for protecting the south coast of England from French invaders, circa 1520 – 1545. The more advanced 18th century methods of cannon casting are graphically illustrated in Diderot's *Encyclopedia* ([1751-75] 1959).

Planish: Hammering of metal to smooth, harden, and strengthen it, as in late medieval cold-hammered armor.

Plastic deformation: Deformation in metals that results from slippage, gliding, twinning, cleavage, or creep and is accompanied by three modes of lattice reorientation. For example, the propensity of the crystal structure of ferrite, when in the form of wrought iron (<0.08% carbon content) to allow slippage, gives wrought iron its soft ductile plasticity. Plastic deformation is the movement of crystal lamellae along slip planes in a particular direction determined by the crystal structure of the lattice network. See lattice reorientation.

Plasticity: In ferrous metallurgy, the propensity of heated crucible steel to be ductile; i.e. to have high workability and to be easily rolled and shaped. No other form of steel has the extreme plasticity of heated crucible steel. The same characterization applies to wrought iron with a carbon content of 0.08% or less. Its silicon content enhances its plasticity and workability. See superplasticity.

Plating mills: North American colonial era ironworks first established in the early 18th century for making sheet iron to produce iron pans and utensils, as well as hoops for coopers and tin plate for whitesmiths.

Polyhedral: A many-sided structure with three or more planes meeting at a point and usually containing more than six plane faces.

Polymetalic societies: Pyrotechnic communities that smelted multiple types of metals, including copper, silver, gold, bronze, and iron. Polymetalism is one of the characteristics of industrial society.

Popham Colony: Location of the ill-fated attempt to establish the "North Virginia" colony, which ended up at the mouth of Maine's Kennebec River, 1607-1608, and the location of New England's first documented shipsmith forge. Archaeological explorations by J. P. Brain located a forge site with associated slag. Also recovered were a ships' caulking iron and strips of wrought iron roves, which functioned as washers, used in conjunction with iron rivets and/or spikes to fasten the planking of the pinnace *Virginia*, the first ship documented as being built in Maine. The roves, spikes, and rivets were almost certainly made on site. The caulking iron was probably a German steel tool brought from England but conceivably could have been made on site from iron bar stock

120

brought by the Popham settlers as ballast.

Post-apocalypse: The unknowable future that will follow the end of the Age of Biocatastrophe when much smaller human populations will have to cope with a radically altered biosphere with dramatically diminished natural resources. Will the techno-elite wave flat world magic wands to mitigate the tragedy of our round world commons?

Post-apocalypse society: A term appropriate for describing social conditions in communities following biocatastrophe and the partial or complete collapse of the industrial infrastructure and the health, social, and public services associated with modern society. In the post-apocalypse, self-sufficient agrarian communities using sustainable technologies will survive along with pockets of crypto-fascist, technologically elite, fortified settlements with the capacity to hoard scarce resources and maintain independent electronic grids. Communities that survive the impact of biocatastrophe on the infrastructure of modern pyrotechnical, industrial society will face the challenge of a decentralized, dysfunctional, post-industrial society unable to provide adequate supplies of fresh water or healthy food or consistently maintain essential services for the majority of survivors. In post-apocalypse society, renewable energy sources will only be sufficient to provide the energy needs of a small component of the once rapidly expanding populations of the world's pyrotechnic global industrial economies. See apocalypse, biocatastrophe, mercury, and infrastructure collapse.

Pouled Janherder: The Persian term for Wootz steel, called "bulat" (steel) in Russian. See bulat and Wootz steel.

Powdered metallurgy: With respect to steel production, the modern technology of utilizing finely powdered iron and selected alloys and molding them in hydraulic die-presses operating under high pressure at high temperatures and for long durations. The end product is steel with a microstructural homogeneity equal to or greater than crucible steel. Due to their alloy content, many powdered metallurgy steel products have characteristics of super hardness and extreme durability. Powdered metallurgy technology bypasses many messy solid/liquid phase transformation processes characteristic of traditional smelting procedures by using whole body melting technology in alloying operations. See superplasticity.

Precision casting: The precise casting and boring of cast iron steam engine cylinders, which required the development of the steam engine in England in the 18th century to become a principal source of power for mills and blast furnaces. First pioneered by John Wilkinson as an improved method of boring cannon, Wilkinson's precision ground cylinders were quickly adapted by Bolton and Watt for production of steam engines. See Wilkinson, John.

Pressed steel: Steel plate bent or pressed by means of dies into channel or other sectional forms, giving great strength with a minimum weight of metal (Audel 1948, 419).

Prime mover: The fundamental power source for the operation of any tool or machine – wind, water, human, animal, biomass, nuclear, or digital. The hand is the primordial prime mover. The steam engine as a prime mover utilizes heat from the combustion of biomass to do work. The utilization of biomass-powered steam engines in England in lieu of water-powered machines was the key innovation that gave birth to an industrial society of textile mills, machine-made fire arms and tools, and steam-powered railroads and watercraft. Ironically, America's extensive river resources allowed the growth of an American factory system that initially relied on water power as its prime mover. The current reliance on non-renewable biomass energy resources as the principal power sources for the heat engines that power our pyrotechnic society is only possible because of the frenzied increase in the consumption of coal, oil, and gas. Costly future sources of nonrenewable biomass fuel resources can be expected to be available with the help of high technology for generations to come. Prime movers can also be defined as sources of technological and thus social change: blast furnaces, printing presses, steam engines, railroads, gasoline, microchips, fiber optics, iPads, computerized high speed trading, CNC technology. The current sudden widespread availability of hydrofracked gas and oil has postponed the end of the Age of Nonrenewable Fossil Fuels but exacerbated the inevitability of cataclysmic climate change and the world water crisis. See biomass fuels, heat engine, Hunter (1979), and hydrofracking.

Principio: The largest integrated ironworks in the Mid-Atlantic colonies. Principio (est. 1731) was located at the head of Chesapeake Bay and was a focal point of the growth of the American iron industry in the Pennsylvania/Maryland region after 1720 (Gordon 1996). Ship fittings produced at Principio and nearby furnaces "ironed" the famous Baltimore clippers constructed in Chesapeake Bay in the century following the establishment of this ironworks. See integrated ironworks and shipsmith.

Privilege: The granting of the right to use the water of a running stream or brook to power a waterwheel for the operation of a trip hammer at a bloomery, finery, or integrated iron works or for powering saw mills, grist mills or water-powered machinery. Individual communities in colonial New England issued privileges from the early colonial period until the middle of the 19th century. See overshot water wheel and trip hammer.

Proeutectiod: The designation for steel, in the form of austenite, formed by ferrite and cementite above the eutectoid temperature of 723° C. When cooled slowly (below 723° C) austenite becomes pearlite, the main microstructural component of cementation steel, with a heterogeneous carbon distribution. If rapidly cooled, martensite is formed. See bainite, hypereutectoid steel, and hypoeutectoid steel.

Proeutectoid cementite: Iron formed upon the cooling of austenite, when carbon in iron is in excess of the eutectoid (0.83% cc). After the formation of pearlite, proeutectoid cementite is deposited at the grain boundaries of pearlite. The microstructures of pearlite and proeutectoid cementite are hypereutectoid (Pollack 1977, 126).

Proeutectoid ferrite: The free ferrite remaining after the formation of pearlite, after the cooling of austenite below the eutectoid temperature. Proeutectoid ferrite is a mixture of ferrite and cementite (0.83% carbon content). The resulting structures, pearlite and proeutectoid ferrite, are hypoeutectoid (Pollack 1977, 126).

Proeutectoid ferrite reaction: "Of the several diffusional reactions by which Austenite can decompose—Proeutectite Ferrite, Proeutectite Cementite, Pearlite and Bainite—the Proeutectoid Ferrite is the most important. This is the first transformation to occur over wide ranges of temperature and composition in most steels produced in large quantities. The kinetics of this reaction thus determine the hardenability of these steels, and at lower cooling rates play a major role in establishing their mechanical properties. The Proeutectoid Ferrite reaction is also the simplest from the viewpoint of ascertaining the effects of alloying elements upon the fundamental quantities determining the kinetics of a diffusional transformation, namely the rates of nucleation and rates of growth" (Kinsman 1967, 39).

Proeutectoid reaction: The response to the impact of cooling rates on the formation of the lamellar structures of proeutectoid cementite and ferrite. "The higher the rate of cooling, the smaller the amounts of proeutectoid ferrite or carbide to be found. The proeutectoid reaction is essentially one of diffusion [and is a function of temperature]. The greater the degree of undercooling the less the size of the ferrite or carbide masses formed" (Bain 1939, 31).

Profiler: A machine for cutting complex dies in the die-sinking process.

Progeny of ironmongers: In the 21st century, not only mechanics, machinists, plumbers, and other tool wielders employed by the palisaded elite, but silversmiths, jewelers, sculptors, printmakers, artists, and artisans of every description, not to mention a vibrant community of blacksmiths, knife-makers, and edge toolmakers.

Psychic numbing: What happens to individuals whose attention is captured by flat world hyper-digital technology, computers, iPods, flat screen TVs, Facebook, etc. Alternatives to the psychic numbing of flat world technology include meditation, nurture, nature (sustainable agriculture, environmental awareness), and the phenomenology of tools, i.e. the marriage of tools, art, and history. The irony of psychic numbing is that flat world technologies have the useful function of enlarging our understanding of our round world biosphere and the finite resources that limit the boundaries of technological innovation and invention.

Puddled bar: The first bar rolled from a puddled bloom of wrought iron after shingling in the decarburization of cast iron; also called a muck bar due to entrained impurities. The puddled bar is often subject to further fining and processing. See finery, muck bar, and puddling furnace.

Puddled steel: Steel manufactured by puddling by the premature halting of the decarburizing process. "Carbon removal is slowed by adding manganese oxide instead of iron oxide, stopping the boil of the puddled iron. Puddled steel is high in slag content and became an important source of steel in Germany and England between 1855 and 1870" (Barraclough 1984a). Barraclough also notes that the oxidation of the carbon was retarded by the "smoky, reducing atmosphere in the furnace during the latter part of the process" (Barraclough 1984b, 93-4). To some extent, puddled steel replaced the traditional production of German steel, which was made by decarburizing cast iron in shaft furnace refineries. In this sense, puddled steel is "German steel" produced in a refractory (puddling) furnace. Barraclough (1984) also notes that "it provided for the needs of industry for some time, in default of a better method, and was capable of producing satisfactory steel from phosphoric pig iron, which is more than could be said for either Bessemer or open-hearth steel up until 1880. It is strange that it has been largely overlooked by almost all of the historians of the iron and steel industry" (Barraclough, 1984b, 93-4). The extent of production in the United States is unknown, but puddled steel may have been used for small scale edge tool production in New England and elsewhere. Halting the decarburization of cast iron to produce malleable iron or steel instead of wrought iron was very difficult due to the challenge of estimating the carbon content of the malleable iron or steel being produced. Unknown is the extent to which individual New England ironmongers mastered this technique and thus supplemented the traditional process of using cementation furnaces to produce blister steel, which then had to be piled and reforged to produce higher quality edge tool steels. See puddling and puddling furnace.

Puddled wrought iron: Iron containing less siliceous slag than direct process bloomery wrought iron and made in reverbatory (puddling) furnaces after 1785 by the indirect process from decarburized cast iron. The development of the puddling or refractory furnace allowed the more exacting production of large quantities of wrought iron with a uniform carbon and slag content. The mass production of this high quality wrought iron and growing production of cast iron produced the iron used in the rapid growth of America's industrial and transportation systems before bulk-processed low carbon steel became available after 1875. Aston (1939) details the wide use of wrought iron, for example, in 19th century bridge construction. Due to the disappearance of both the bloomery and the puddling furnace by 1930, wrought iron was no longer produced by these long established techniques in any significant quantities in the U.S. See furnace types, puddled steel, puddling furnace, reverbatory furnace, and wrought iron.

Puddling: The process of decarburizing cast iron and oxidizing its slag constituents in a reverbatory furnace. The resultant products include puddled steel, malleable iron, and wrought iron.

Puddling furnace: A redesigned version of older refractory furnaces, also called the reverbatory furnace. The first of Henry Cort's puddling furnaces was built in 1784 and was one of the most important advances in iron and steel production during the second stage of the Industrial Revolution. The puddling furnace kept the fuel separated from the ore to avoid recarburization and was used for producing relatively slag-free wrought iron from pig iron, using coal or coke as a fuel. A secondary function of the puddling furnace was the production of malleable iron and, after 1830, puddled steel. See furnace types and refractory furnace.

Pyrotechnic pie chart: The legacy of the pyrotechnic society including its many ironmongers and swordsmiths can be expressed as a pie chart of ecotoxins and biologically significant chemical fallout (BSCF). CO_2 emissions (the carbon footprint) is only one small slice in a pie chart that includes methylmercury and other heavy metals, dioxins, perfluorinated chemicals, and chlorinated hydrocarbons of all kinds (organochlorides, polychlorinated biphenyls [PCBs], phthalates, polybrominated diphenyl ethers [PBDEs], bisphenol-A, etc., many of which have dozens of congeners [subspecies]). Anthropogenic radioactivity is also one slice of the pyrotechnic pie chart. Every rain shower is a chemical fallout event. See biocatastrophe and tragedy of the world commons.

Pyrotechnological products: The glass, pottery, lime (cement), copper, bronze, iron, steel, and tool steel alloy products produced by biomass-consuming polymetalic pyrotechnic societies from the Copper Age to the contemporary era. Environmentally significant byproducts of polymetalic industrial societies include carbon dioxide (CO_2) and biologically significant chemical fallout, including numerous persistent organic pollutants (POPs), methylmercury, and long-lived radioisotopes. See apocalypse, mercury, phenomenology of tools, post-apocalypse society, biocatastrophe, and infrastructure collapse.

Pyrotechnology: The process of using fire to make glass, terra cotta (pottery), cement (lime), iron, and other metals. Pyrotechnology is the primordial industrial basis for civilized society. Given the environmental consequences of pyrotechnic societies, (including climate change, resource depletion, and biologically significant chemical fallout) after five millennia of the growth and evolution of pyrotechnology and 250 years of the massive productivity and emissions of a full blown Industrial Revolution, pyrotechnological consumer societies may be sailing close to the wind, i.e. approaching a rubicon, after which biocatastrophe is inevitable. See biologically significant chemical fallout and mercury.

Quants: Mathematical wizards, often with PhDs from Harvard, who operate the high speed electronic trading system that makes the shadow banking network the most profitable of all endeavors of our growing klepto-plutocracy.

Quench hardening: Making steel more homogenous by sudden cooling for the purpose of hardening. Hardened steel edge tools were tempered (reheated to less than the forging temperature and slowly cooled to relieve brittleness). See austenite, martensite, quench, and temper.

Quench: To cool steel rapidly in order to increase its strength and hardness. Quenching also results in extreme brittleness of steel, requiring further heat treatment (tempering). Brine, cold water, and oil are the fastest quenching mediums, respectively, for austenized steel. Air is the slowest cooling medium. Ancient quenching techniques included the use of urine and other assorted odd confabulations. See carbon content of iron and steel, martensite, quenching threshold, and temper.

Quenching method: Method that determines hardness as a function of the rate of cooling. Brine, water, oil, and air are the most common quenching mediums. Oil hardened steels deform less than water quenched steels; air hardened steels deform the least.

Quenching temperature: Temperature required for the rapid cooling of steel. Higher carbon content steel requires a lower temperature for efficient quenching; lower carbon content steel requires a higher temperature.

Quenching threshold: The point at which iron containing less than 0.5% carbon cannot be significantly hardened by quenching and, therefore, cannot be made into tool steel or quality edge tools. A large percentage of modern steel is low carbon steel (0.2 - 0.5% cc), which can only be slightly hardened by quenching.

Question for the reader: Will clever techno-elite entrepreneurs and bioengineers create magical flat world solutions for the unfolding tragedy of the round world commons?

REM: Reflection electron microscope. Along with the transmission electron microscope and scanning electron microscope, REMs are important tools used for the study of the microstructure of metals. The HRTEM is the most advanced form of the resolution electron microscope. See electron microscope, HRTEM, SEM, and TEM.

R. M. S. (Robert Mushet steel): Special self-hardening steel introduced by Robert Mushet in 1868. The forerunner of high speed steel, R. M. S. contained 7% tungsten, 2% manganese, and increased metal cutting speeds by 50% (Tweedale 1986). One important advantage of R. M. S. was that it hardened upon being air-cooled, whereas, most steels had to be quenched for hardening. Later, it was discovered that manganese along with tungsten was responsible for its increased efficiency in comparison to high carbon tool steel without alloy additions. See Mushet, Robert F. and Taylor-White.

Rabble: The tool used by the iron puddler in a reverbatory furnace to stir, form, and remove the loup of wrought iron in the furnace.

Radanites: Jewish traders operating along the Silk Route, 500 AD - 1200 AD, who brought Wootz steel from India to Persia for Damascus sword production.

Reaganomics: The radical philosophy promoting the crypto-fascist ideals of a self-indulgent unfettered free enterprise system manifested in the growing income disparities of a flourishing klepto-plutocracy and its shadow banking network. The financial crisis of 2008 is a recent expression of a growing threat to sustainable economies by a philosophy of greed, which ironically undermines our traditional free enterprise system by empowering a few at the expense of a vast underclass of victims of financial exploitation. Its antithesis is invisible sustainable economies where hand tools are the symbol of self-sufficiency and viable alternative lifestyles.

Reagent: With respect to Damascus steel, the acidic medium, such as lemon juice, used to reveal the particular crystal structure that results from the forge welding of the sword or gun barrel by a particular smith.

Rebar: Round steel bar stock commonly used for bridge and building construction. In addition to its main constituent of pearlitic steel, a typical sample of rebar has the following chemical composition: carbon +/- 0.45%, manganese 0.80 to 1.0%, silicon 0.2%, copper 0.3%, nickel 0.08%, chromium 0.1 to 0.2%, sulfur 0.04 to 0.05%, molybdenum 0.04%, and phosphorus 0.01 to 0.02%. The complex chemical composition of this commonplace steel product illustrates the ubiquitous presence of important alloys, such as manganese, silicon, copper, nickel, and molybdenum as microconstituents in steel products (NUCOR 2007).

Recalescence: The sudden reappearance of glowing light in metal being cooled, which occurs during the transition of beta iron to alpha iron, which also gives off heat.

Recarburizer: Carbonaceous material, including pig iron, added to molten steel to increase its carbon content. Recarburization is analogous to the ancient process of submerging wrought iron in a bath of pig iron to produce Brescian steel.

Recovery: Loss of lattice dislocation through the elimination by diffusion of strain hardening from thermal stress and the subsequent re-crystallization of the lattice structure of a metal. Recovery can occur as a consequence of tempering but also may occur gradually at room temperatures.

Re-crystallization: Process resulting from the use of heat treatment to form new crystal lattice structures in deformed metals.

Recycled steel: The product of a common strategy for obtaining steel for hand tool production before the era of bulk process steel production, such as the reuse of dull steel files and rasps, which were commonly reforged into knives, turning tools, farrier's tools, and other implements requiring a high carbon content. This strategy for reclaiming steel, commonly used by the blacksmith, was the precursor of the later use of steel scrap as part of the charge used to make crucible and open hearth steels. See reforged steel.

Red brass: Brass consisting of 85% copper, 5% tin, 5% lead, and 5% zinc.

Red hardness: Term for the capacity of cutting tools to operate at temperatures above 900° F.

Red heat: 650° - 850° C, the temperature at which the spheroidization of cementite by mechanical treatments (hand forging, drop-forging, or rolling) can occur.

Red ocher: Finely ground hematite. The mineral iron oxide is the most valuable of all iron ores (Fe_2O_3). Red ocher is perhaps best known for its ceremonial uses in indigenous Native American burial sites.

Red short: "When impurities, such as sulphur arsenic, render the metal unworkable at a red heat, it is said to be *red short*" (Spring 1917). The red short metal's brittleness makes it subject to shattering when hammered or forged.

Reduce: To transform an element by the removal of oxygen from a highly oxidized state to a less oxidized state, as in the conversion of iron oxide to iron in the smelting process.

Reducing slag: Slag that facilitates the removal of oxygen. See slag.

Reductant: The reactant, e.g. carbon monoxide, which facilitates the chemical reduction of iron oxide to iron, also producing carbon dioxide while removing the oxygen.

Reduction: The partial or complete removal of oxygen from an oxide. See oxidation reduction reaction.

Refined iron: **1)** In a generic sense, any rolled or hammered bar stock produced by a direct process bloomery or a blast furnace finery, subject to additional mechanical or thermal treatment (refining) in a finery to further remove slag, add or remove carbon, and more closely align slag inclusions. Hence the terms "double refined" and "triple refined" descriptions of refined iron that has undergone additional thermal and mechanical treatment. See finery and refinery. **2)** The result of an optional intermediate step in the production of fine quality wrought iron in the puddling furnace, whereby pig iron is partially decarburized in a run out (cupola) furnace in which multiple tuyères produce an air blast that removes much of the silicon and carbon in the pig iron. The iron is then cast in iron molds and cooled by water circulating in the hollow walls of the mold. The resulting iron, in the form of plates or ingots, is then broken up and further decarburized in the puddling furnace to produce high quality wrought iron. See cupola furnace.

Refinery: "Hearth used to remove silicon from pig iron preparatory to fining or puddling, usually fired with coke" (Gordon 1996, 310); not to be confused with "finery," an open-hearth furnace used since ancient times to refine either bloomery iron or cast iron. The refinery represented an additional step in the industrial process utilized to prepare the cast iron for subsequent manufacturing processes, such as re-melting, rolling, or drop-forging, especially in the 19th century. See finery.

Reflection electron microscope: See REM.

Reforged steel: A common source of tool steel prior to the era of bulk-processed steel < 1870. Typically, worn out steel rasps and files were reforged by heating and hammering and thus recycled to make edge tools, farriers' tools, and other steel implements. Reforged rasps and files often exhibit vestigial traces of the teeth of the original tool. See recycled steel.

Refractory: In the modern era, any high temperature resistant material used to line furnaces, i.e. "refractory" metals or ceramics. Refractory is another term for the crucible used in cast steel production, circa 1800, which necessitated the use of special high temperature resistant clays (+/-1500° C). The principal contribution of Benjamin Huntsman's reintroduction of the crucible steel process (1742) was his use of high temperature resistant Stourbridge clays to manufacture his crucibles. The unavailability of such refractory clays in America postponed the advent of cast steel production until 1850 and the introduction of graphite crucibles. See Stourbridge clays and crucibles.

Refractory furnace: Another name for the cementation furnace, which operated through the use of sealed refractory chambers to produce blister steel where there was no fuel to ore contact. See blister steel, cementation furnace, puddling furnace, and puddled steel.

Regenerative gas furnace: A form of furnace that recycles and burns waste blast furnace gases, further increasing blast furnace efficiency. The regenerative gas furnace was introduced in America in 1857, shortly after Sir Carl Wilhelm Siemens developed it in France.

Regeneration: The use of waste heat and escaping gases to preheat incoming air in blast and other furnaces.

Renewable energy resources: Energy sources that are both finite and continuously available due to the biogeochemical cycles of the chemosphere, e.g. hydroelectric, solar, geothermal, tidal, biomass, and wind. There are sufficient renewable energy resources to sustain sophisticated industrial economies with a population of +/- two billion people, who are supported by sustainable agrarian and ocean fishery communities of another +/- one billion persons.

Rennwerk: The direct process reduction of iron ore to wrought iron, in which steel is sometimes a byproduct. In Sweden, circa 1600, the production of natural steel as a byproduct of Rennwerk was one of three options for making steel. The other two were fining (decarburizing) cast iron (also called German steel) and carburizing wrought iron in a steel furnace (blister steel). See German steel, blister steel, and puddled steel.

Repoussé: The formation of raised designs by hammering the reverse side of malleable sheet metals being utilized for jewelry (etc.) design and manufacture. Combined with chasing, the technique is utilized to create three-dimensional works of art. See chasing.

Resistance to decarburization: Property of iron being smelted to not lose carbon due to furnace condition and slag content. Resistance to decarburization varies with the alloy content of the steel (e.g. chromium).

Retained austenite: Austenite retained as a result of adding alloys to carbon steel, which prevents the transformation of austenite to martensite, allowing as much as 35% of the structure of steel to remain as non-magnetic austenite, e.g. as in stainless steels. Tempering will eliminate most retained austenite (Shrager 1949).

Retort reduction process: A modern process for producing sponge iron by using preheated natural gas injected into a furnace retort and forced through the furnace charge. This process begins with charging and is followed by primary reduction, secondary reduction, and cooling (Pollack 1977). See sponge iron.

Reverbatory furnace: "Furnace with a brick roof heated by flame. Metal on the hearth is heated by reflection and radiation from the roof and does not come in contact with the fuel" (Gordon 1996, 310). In the reverbatory furnace, both the hearth and the roof are built from refractory brick, and the flame enters the furnace from the side. Henry Cort's improved design of the reverbatory furnace in 1784 was a key event in the Industrial Revolution, enabling the rapid production of larger quantities of wrought and malleable iron for hand-forged hardware and tools. The reverbatory furnace was especially important for the refining of cast iron into nearly pure wrought iron to be used in the production of cementation steel, which was then utilized in producing pure crucible steel (cast steel). See crucible and puddling furnace.

Riddle: A foundry sieve usually made of brass or galvanized iron cloth for removing solid materials from molding sand.

Roasting oven: Oven commonly used before 1850 to preheat and dry iron ore before smelting in either a bloomery or blast furnace. The roasting process was a variation of the ore dressing process, which could also be done by manual crushing with a sledge and air drying. The objective of these processes was to make the surface of the iron ore more porous and, thus, susceptible to the effects of reducing gases in smelting processes.

Rockwell hardness test: "A method of determining the hardness of metals by indenting them with a hard steel ball or a diamond cone under a specified load, measuring the depth of penetration, and subtracting the latter from an arbitrary constant. Rockwell hardness numbers are based on the difference between the depths of penetration at major and minor load; the greater this difference, the less the hardness number" (Shrager 1949).

Rod mill: A mill utilized for rolling billets into semi-finished hot-rolled rods used for wire drawing. See billet.

Rolling direction: The direction of rolling that determines the grain direction in steel or iron bar stock. For knife and edge toolmakers cutting curves or patterns across the grain, reforging the grain direction is necessary before a new directional pattern in the lattice structure of the steel or forged iron tool can be created. See rolling mill.

Rolling mill: Mill used after the decarburization of cast iron in a finery or further refinement of the anconies produced in a chafery to shape malleable iron bar stock into useable forms, such as nail and wire rods and sheet iron, for forging into nails, tools, utensils, and hardware by a blacksmith or shipsmith, or, later, into rails (1839) and I-beams (> 1870). Rolling removed impurities and further strengthened the iron or steel being rolled. In 1784, Henry Cort designed the basic form of the modern rolling mill, adding groove rollers to rapidly shape the puddled iron produced by his recently redesigned puddling furnace into the multiplicity of forms suitable for manufacturing ironware of every description. Rolling mills that produced sheet iron and steel for armorers appeared at least as early as the 16th century, during the height of the German Renaissance, and were used by the mid-18th century to produce sheet silver for silversmiths. Rolling mills are a ubiquitous component of all modern (>1870) bulk process steel producing communities. Modern primary rolling mills, similar to or synonymous with chaferies, use pit-type furnaces to soak steel ingots before hot rolling into semi-finished products such as bar stock, slabs, billets, sheet iron, and blooms of pure steel. A variety of secondary rolling mills further process them into finished products. Steel ingots utilized in primary rolling mills often weigh in the range of 10 to 40 tons or higher. See cast iron, chafery, decarburization, finery, forge, furnace types, malleable iron, bloomery, and puddling.

Round hearth: Innovative hearth design introduced in America in 1832 that increased the efficiency of blast and bloomery furnaces.

Round world: The biosphere and its inventory of finite resources. The context for the history of ironmongering ; the iconography of the tools thus produced tell us about the history of technology: cascading Industrial Revolutions eventually evolve into to the Age of Biocatastrophe as finite round world natural resources are depleted or contaminated.

Ruff, Drawn, Gadd, and Slick (steel): Early 18th century English terminology described by Barraclough (1984a) pertaining to cementation steel (ruff) and its faggotted and welded enhancements, i.e. spring (drawn), single shear (gadd), and double shear (slick). German immigrants living in or near Newcastle (1680 through 1730) were instrumental in introducing continental steel finishing strategies and applying them to the production of improved varieties of cementation steel. See blister steel and cementation steel.

Rule of thumb: A term used to describe work patterns based on the empirical experience of generations of ironmongers. Prior to the development of modern chemistry in the 19th century, blacksmiths and edge toolmakers had no knowledge of the chemistry of steel- and toolmaking strategies and techniques. The role of carbon in steelmaking was unknown. Significant alloys such as manganese and silicon, which played such an important role in determining the characteristics of iron and steel, were as yet unidentified. Nonetheless, high quality edge tools were produced over long periods of time from the early Iron Age to the early modern period by these repetitive, intuitive work patterns (Barraclough 1984a). See kan of the ironmonger.

Russian iron: Sheet iron with a very hard and highly polished surface, made in Russia by a secret process and used for steam engine boilers, etc.; also known as planished iron due to its smooth surface. See planish.

SEM: Scanning electron microscope. Along with the reflection electron microscope (REM) and the transmission electron microscope (TEM), SEMs are commonly used analytical tools for analyzing the microstructure of metals by detecting low energy secondary electrons emitted from metals due to excitation by a primary electron beam. Detectors then map the data by noting beam position. See electron microscope, HRTEM, TEM, and REM.

Sagger: Metal pot used to hold castings to be malleableized by annealing.

Salamander: Slab or bear of solidified cast iron produced by high shaft direct process bloomery furnaces as unwanted waste products during the bloom smelting process. In central Europe, these salamanders may have been the basis for recognizing the possibility of decarburizing cast iron to produce German steel by altering the fuel to ore ratio in the smelting process. Many salamanders were later recycled in blast furnaces due to their high iron silicate content. See direct process, iron silicate, and slag.

Salt bath: Baths of molten chemical salts utilized for hardening and tempering a metal.

They provide uniformity of heat treatment as well as protection from oxidation.

Samurai sword: See Japanese samurai sword.

Samurai/Damascus sword surface pattern: Pattern produced by the hardening of cementite grains during the forging process and shown as layers of bright areas in blade cross sections after blade surface treatment by reagents. See damascene and Damascus sword.

Sand mold: Mold made with wooden patterns that allowed precise shaping of cast iron products, first used by Abraham Darby at his Bristol iron foundry in the early 18th century. Sand molds replaced the earlier technique of molding with hay, clay, and loam and could be utilized for multiple castings. Wooden patterns and sand molds paved the way for the production of Newcomen steam engine cylinders by Abraham Darby II (c. 1722-1733). See cast iron, foundry, and molding.

Sandwich technique: One of several ancient strategies for making knives by which sheet iron is wrapped around a steel core, which is then ground away to expose the cutting edge Pleiner (1962).

Sap: Traditional term for the characteristic appearance of the fracture of blister steel produced in cementation furnaces prior to any further thermal or mechanical treatments. Changes in the appearance of sap express the changes in the crystal structure of iron in its conversion from malleable iron to steel. The appearance of the fracture of blister steel after smelting was the key to determining its carbon content. During the conversion of bar iron to blister steel in the converting furnace, "sap" appeared to signify ongoing changes in the crystal structure in the bar iron ("a rim of fine crystals around an unaltered coarse central structure" [Day 1991, 270]). The sap, i.e. coarse crystals indicating blister steel with a low carbon content (0.5 – 0.6% cc), was gradually changed into a fine crystalline pattern and the percentage of the sap in the fracture became lower. According to the Sheffield classification of blister steel, the sap entirely disappeared in steel having a carbon content of 1.15% and was replaced by a pattern of fine crystals signifying fully austenized steel. At a carbon content of 1.4 and above, coarse crystals began reappearing, signaling increased cementite content (Barraclough 1984a, 44-5). See cementation furnace, fracture, and Sheffield classification of blister steel.

Saugus Ironworks: America's first integrated ironworks established on the Saugus River in Saugus, MA, in 1646. This ironworks, which was also known as Hammersmith, included a blast furnace, finery, chafery, and a blacksmith shop, which was used by Joseph Jenks (later Jencks) for scythe and edge tool production. The flux used at this blast furnace was a unique form of limestone called grabbo, brought by boat from nearby Nahant. The Saugus Ironworks also functioned as one of New England's first known shipsmith forges.

Satetsu: The iron sand used by samurai sword makers to produce a raw steel bloom (kera) in a direct process open-hearth clay furnace (tatara) prior to the final forging of the sword. See bloom, breakdown furnace, and tatara.

Scanning electron microscope: See SEM.

Scarf weld: An ancient technique for making knives and swords that involves the steel being laid on cold iron and then wired, heated, and welded to create a steeled edge tool. Also, see Bealer (1976) for a contemporary description of the blacksmith's art of scarf welding, i.e. the hammering and welding of two overlapped pieces of metal. See forging.

Secondary hardness: Phenomenon of the resumption of hardening that occurs during the tempering of high alloy steel after the initial quench hardening.

Self-hard: The first commercial alloy steel, R. M. S., invented by R. F. Mushet in 1868 in his Forest of Dean ironworks. He added 8% tungsten to the molten steel, allowing it to harden in the air with no rapid quenching in water. See Mushet, Robert F. and R. M. S. steel.

Self-hardening steel: High carbon alloy steel, which hardens at room temperature without oil or water quenching. Its alloy content slows the transformation of austenite to pearlite.

Semi-steel: High-grade grey cast iron, which is made stronger by the addition of steel scrap in the charge and contains lower silicon, phosphorous, and total carbon content than grey cast irons (3.2% versus 3.4% for grey cast iron and 3.45% for soft grey cast iron). In contrast, the carbon content of white cast iron and annealed malleable cast iron is 2.75%. Many of the pieces of machinery on display at London's Victoria and Albert Science and Technology Museum made between 1800 and 1840 may be semi-steel. See carbon content of iron and steel and cast iron.

Shadow banking network (SBN): Along with multinational corporations, the SBN is a significant component of the klepto-plutocracy that dominates American economic activity in the 21st century. Based on high speed trading, the sale of collateralized debt obligations and credit default swaps, currency manipulations, and parasitic hedge funds, the SBN is dependent on computerized mathematical formulas concocted by "quants", e.g. Harvard PhDs with a doctorate in math. While traditional banking institutions provided essential services to local businesses and homeowners, the SBN utilizes hundreds of trillions of dollars of debt as "bank assets," which are constantly traded in a world banking network "casino". The resulting commissions and profits, which derive from the manipulation of gross world assets, are collected by a tiny percentage of individuals with extraordinary incomes, i.e. the klepto-plutocracy.

Shaft furnace: Common form of open-hearth furnace for smelting iron throughout the early Iron Age, along with the bowl furnace. Shaft furnaces evolved into blast furnaces as the length of the shaft grew in size. Early Iron Age shaft furnaces were difficult to control and often produced iron with a heterogeneous carbon content, ranging from liquid cast iron and natural steel to malleable iron, though low carbon wrought iron might have been the goal of the smelt. The high shaft bloomery "Stucköfen" furnace of south Germany was the final stage in direct process iron production (circa 1500 AD). The larger shaft furnaces which followed were true blast furnaces producing cast iron that needed refinement into wrought iron and/or steel. See bloomery, bowl furnace, German steel, open-hearth furnace, slag pit furnace, and slag-tapping furnace.

Shaft hole vs. Socket hole: Two alternative methods of securing handles to adzes, axes, and hammers. In Neolithic Europe, socket hole tool design was "the traditional method of hafting the head to a knee-shaped, cleft handle. In the course of time, the shape of the head was drastically modified, yet even when casting was adopted they still kept to the old way of lashing the head to the shaft" (Goodman, 1964, 14). In the early Iron Age in Europe, some socket holes were of square design; "It was about this time, between 500 and 200 B.C., that this traditional method was finally abandoned, and the ax assumed the form familiar to us today, a heavy wedge-shaped forging, with the shaft-hole, usually elliptical in cross-section, made by folding a rectangular billet of iron in the middle, leaving the eye parallel to the cutting edge" (Goodman, 1964, 20). Goodman illustrates shaft hole axes and adzes dating as early as 2900 BC, including Mesopotamian tools utilizing this modern form. One of the mysteries of history is why primitive socket hole tool forms continued to be used in Cis-Alpine Europe... "One of the most curious blind alleys in the whole checkered history of human progress" (Goodman, 1964, 14). See Goodman (1964), Childe (1925), and Brack (2008).

Shear steel (Sheaf steel): Steel resulting from the reworking of blister steel to increase purity and homogeneity by taking cut bars of blister steel and piling, reheating, re-rolling, re-hammering, and re-welding them; also called "spring" steel when rolled into sheets for saw-makers. The mechanical working of bundled bars of blister steel more evenly distributed its carbon content. It is called shear steel "because blades for shears for cropping woolen cloth were always made this way" (Spring 1917, 117). The term shear steel also derives from the simple process of shearing blister steel bar stock before it is repiled and reforged into higher quality steel. Prior to the production of crucible steel, shear or "sheaf" steel was the highest quality steel available. Implements marked "shear" or "sheaf" were usually manufactured before 1784, after which the designation "cast steel" became more commonplace. Many a shipwright's tool was made from shear steel by New England's edge toolmakers and sword cutlers. Cutlers probably used it for swordmaking due to its heterogeneous carbon content, in contrast to the homogenous carbon content of pure cast steel.

Shear stress: The stress threshold for crystal flow (slip and creep). Shear stress in the lattice structure of metals is temperature dependent, decreasing with increasing temperature and reaching zero at the melting point.

Shearing movements: Rapidly occurring microstructural changes typified by the acicular formation of martensite and the twining formations in the intermediate phases of bainite. The slow rate of nucleation, as in the formation of pearlite, is the antithesis of shearing.

Shearing stress: Stress that changes the angles between the faces and the lengths of the diagonals of microstructures, acting tangentially to the surface of contact without changing the lengths of the sides of the structure undergoing stress. "Shearing stress intensities are of equal magnitude on all four faces of an element" (Baumeister 1958, 5-23). The stress threshold for crystal flow (slip and creep) and shearing stress in the lattice structure of metals is temperature-dependent, decreasing with increasing temperature and reaching zero at the melting point. See stress fields, slip, creep, crystal structure of metals, and martensite shearing.

Sheet iron: Iron sheets produced by plating mills beginning in the early 18th century for use as hoops by coopers and tin plate by whitesmiths.

Temper Number	Mean Carbon Content, %	Name	Fracture Appearance
1	0.60-0.70	Spring Heat	80% sap
2	0.75-0.85	Cutlery Heat	60% sap
3	0.90-1.00	Shear Heat	40% sap
4	1.05-1.15	Double Shear Heat	20% sap
5	1.20-1.35	Steel Through Heat	Fine crystals throughout
6	1.40-1.60	Melting Heat	Coarse crystals throughout
7	1.70-2.00	Glazed Heat	Very coarse and facetted crystals

Table 4. (Barraclough 1991, 270) [sap: unaltered coarse central structure]

Sheffield classification of blister steel: The classification of the visual appearance of the unaltered microstructures of the central portion of a piece of blister steel bar stock after manual fracturing. The visual appearance of these microstructural formations was called "sap" by 18th century English steelmakers and based upon the coarseness of the grain structure as shown in Table 4.

Sheffield plate: Fused copper and silver, developed in the late 18th century as imitation silver by Thomas Boulsover (Tweedale 1987). Sheffield plate is usually in the form of sheet copper sandwiched between two thin layers of silver.

Shiage: The rough grinding and shaping of both sides of a sword blade to be tempered using a two-handled drawknife made from a steel sword. Shiage is the first step in the heat treatment of the Japanese sword. See Japanese samurai sword, naginata, tsuchioki, yaki-ire, and yaki-modoshi.

Shingler: The manipulator of the puddled ball being hammered at a puddling furnace.

Shingle the loup: The first step in manufacturing wrought and malleable iron bar stock. The forge master hammers the hot metal bloom of wrought iron to expel the impurities, shape the bloom, and align the silicon fibers remaining in the bloom to enhance the tensile strength of the metal. Shingling preceded the fining (reheating, re-hammering) of wrought or malleable iron, which was then shaped into bars called anconies. See merchant bar, piled wrought iron, puddled iron, refined iron, and rolling direction.

Shipsmith: A blacksmith with a specialty in making both the iron fittings and edge tools needed to construct a sailing ship. Ubiquitous in England, northwestern Europe, and colonial New England before 1800, shipsmiths sometimes also made anchors. Both the North River and the Taunton River in the bog iron country of southeastern Massachusetts were the location of the forges of numerous shipsmiths working between 1650 and 1750. After 1800, trade specialization resulted in shipsmithing and edge toolmaking being regarded as separate trades, as illustrated in the 1856 *Maine Business Directory*. The shipsmiths and edge toolmakers listed in Maine directories lived in coastal shipbuilding communities or nearby upstream towns with water power for trip hammers and easy access to the Maine coast by river. See *Art of the Edge Tool* (Brack 2008), adz, broad ax, and slick.

Siemens-Martin process: Process that enabled the melting of any combination of cast and wrought iron to make steel. This process combined the melting of pig iron in open-hearth furnaces by the Siemens process with the addition of molten pig iron to the charge utilized by the Martin process, which originated in France. In 1865, the Martin brothers utilized the regenerative gas furnace designed by Siemens to develop one of the first bulk steelmaking technologies, which was later perfected in 1877 with the introduction of basic hearth linings. After 1877, the basic open-hearth process produced more tonnage of steel than the acid open-hearth process, which could not utilize iron ores or pig iron high in phosphorous. See open-hearth furnace and Siemens open-hearth furnace.

Siemens open-hearth furnace: Along with the Bessemer process, the most important steel production furnace type from 1870 to the mid-20th century. The Siemens open-hearth furnace replaced the Siemens-Martin pig and scrap furnace, providing better quality control of the steel production process by the addition of carefully measured amounts of iron-oxide-producing iron ore to the slag after the melt was underway. This addition of iron ore during the melt caused spectacular boiling reactions at the slag/metal interface, as a result of the burning of carbon monoxide, also differentiating the Siemens open-hearth furnace from the Siemens-Martin furnace.

Siemens regenerative gas furnace: Gas furnace adapted for open-hearth furnaces as well as for more efficient crucible steel production, both in England and America; developed in 1857.

Silent majority: The well meaning, usually convivial, and often Christian majority of the 90% who are neither Tea Party Taliban nor crypto-fascists. They are an economic underclass whose indirect computer drive exploitation is the source of the ever growing power of the American klepto-plutocracy.

Siliceous slag: The slag or iron silicate content in wrought iron (1.0 - 3.0%). Its content depends on the type and extent of mechanical treatment of iron by the smith with hammer and anvil, water-powered trip hammers, or rolling mills. Most siliceous slag is removed by the blast furnace in indirect iron and steel production but remains in wrought iron after direct process bloomery smelting. After bloomery-smelted wrought iron is refined in a finery furnace, variable amounts of slag remain in the iron, depending on the refining techniques and objectives of the finer. These iron-silicate microconstituents play an important role in enhancing the plasticity of edge tools forged from malleable iron and give older edge tools their convivial "feel" and ease of sharpening in contrast to most "hard" modern edge tools and plane blades. Trace quantities of silicon may also be added to cast steel as the result of chemical reactions in the walls of the clay crucibles, which shed tiny amounts of silicon-bearing material back into the crucible steel. The role of silicon as a microconstituent in the production of fine edge tools has not yet been scientifically documented. See carburize, forge, forge welding, and silicon as a softener.

Silicon: An important deoxidizer during the smelting of iron that softens cast iron and increases its fluidity by facilitating the formation of graphite from carbon, strengthens ferrite by making it more resistant to oxidation, strengthens the microstructure of pearlite by the solid solution hardening of ferrite, decomposes other carbides, and enhances graphite formation. When present in steel in concentrations up to 2%, silicon intensifies the tendency of molybdenum, manganese, and chromium to increase hardenability and enhance carbide formation. It may also soften edge tools, increasing their ductility. See siliceous slag.

Silicon as a softener: The capacity of silicon to help precipitate chemically combined carbon into free graphitic flakes. The higher the silicon content is the more combined carbon becomes graphite. Silicon greatly increases the machinability of cast iron pieces, such as iron valves and hollowware. Silicon also plays a role in making the old cast steel plane blades and edge tools from Sheffield and Pittsburg superior to the tools drop-forged from steel made in the modern electric furnace. Silicon as a microconstituent produces a softer more congenial cutting edge, traditionally preferred by shipwrights, timber framers, and case furniture makers. The silicon content of edge tools may also increase durability by increasing the tensile strength of the cutting edge, relieving the brittleness characteristic of many modern edge tools. See siliceous slag.

Silicon steel: An alloy steel containing silicon. If steel contains 1 to 2% silicon, it is very tough and suitable for automobile springs. With 3 to 5% silicon, it has improved magnetic properties and is often used in electrical appliances.

Silver steel: "A soft and ductile steel with a low carbon content. It is supplied ground to a bright finish in small-diameter rods. It is useful for model makers and in laboratories. It cannot be hardened owing to its low carbon content" (Salaman 1975, 247). Noted by Cyril Smith as "the first alloy steel to become – undeservedly – popular" (Smith 1960, 141). Stodart and Faraday utilized it during their attempts to duplicate Damascus steel. Also noted by Lee (1995) as follows "This mark is commonly found on old handsaws… when it was discovered that adding a percentage of chromium to high carbon steel not only increased the hardness and toughness but also gave the steel a bright silvery finish that improved rust resistance, the alloy was quickly dubbed silver steel" (Lee 1995, 20).

Simple machine: "In physics, a machine is any device that can be used to transmit a force and, in doing so, change its size or direction… simple machines include the inclined plane, the lever, the screw, and the wheel and axle. All of these machines illustrate the concept of work" (Maiklem 1998, 26). The pulley and block and tackle are simple machines. See machine, *Tools Teach: An Iconography of American Hand Tools* (Brack 2013), turbine, and work.

Singer, Nimick & Co.: Established in 1849 in Pittsburgh, by 1853, Singer, Nimick & Co. was the first reliable producer of cast steel in the US (Gordon 1996). It is unknown to what extent this company supplied edge toolmakers, such as Timothy Witherby, the Buck Brothers, and the Underhills, who manufactured cast steel woodworking tools during and after this time.

Sinter: To produce hard solid metal or ceramic artifacts from atomized powders (e.g. iron powder) using hydraulic dies operating at high temperatures and under high pressure. See powdered metallurgy.

Skelp: Strips of mill steel used to produce lap-welded and/or butt-welded steel tubes by drawing the strips through a bell-shaped pattern at welding temperature.

Skilled labor shortage: One of the ironies of the Age of Information Technology and the early years of the Age of Biocatastrophe is a growing skilled labor shortage despite intensifying structural unemployment caused by the IT revolution, CNC technologies, and the increasing use of robots to execute industrial and manual functions. Artisans and craftspersons skilled in the use of hand tools, born-again ironmongers as it were, are increasingly in short supply for maintenance and repair of the infrastructure of a global consumer society that still needs plumbers, carpenters, electricians, mechanics, and communications systems specialists. The widespread deletion of vocational education, a rapid decline in the quality of public education, and a failure to invest in technical training of a contemporary workforce skilled in the use of hand tools engender this shortfall.

Slack quenching: To quench at a rate too slow to produce martensite. When slowly cooled, austenite instead forms pearlite, lamellae of ferrite, and cementite arranged by twinning into Neumann bands. Slack quenching "worked best on a relatively low carbon material. Damascus swords were quenched slowly in oil, or by a current of air" (Smith 1960, Footnote 24). See acid process, basic process, Damascus steel, Damascus sword, ferrosilicate, flux, and manganese.

Slag: The non-metallic waste product of bloomery and blast furnaces, usually a silicate of calcium, magnesium, and aluminum, which covers the molten metal in the smelting process. Varying amounts of iron oxides, especially iron silicate, also characterize slag. The slag of small bloomery and open-hearth furnaces contains more wasted iron in the form of iron silicate than the slag of more efficient blast furnaces, especially if limestone was used as the flux in the blast furnace. The slag is usually skimmed off of the molten metal prior to tapping or pouring. See blast furnace, bloomery, salamander, and siliceous slag.

Slag heap: The siliceous slag-bearing waste products of bloomeries and blast furnaces; the tell-tale evidence of smelting activities. See slag.

Slag pit furnace: A common form of early bowl or shaft furnace. The slag accumulated in pits dug underneath the furnace. High in iron oxide, many slag blocks were recycled in blast furnaces, which more efficiently reduced their iron content. Slag pit furnaces were often moved for re-firing when the slag built up under the firing chamber, resulting in the accumulation of huge quantities of slag in early Iron Age smelting locations, such as the Black Mountains of southern France.

Slag-tapping furnace: A furnace in which slag was tapped by draining liquid slag from the furnace using a trench; the most common form of direct process furnace from the Roman era to the late medieval period (Tylecote 1987).

Slag viscosity: "The viscosity of a fluid is a measure of its resistance to gradual deformation" (Wikipedia 2013). Slag viscosity is increased by low iron oxide content, as in efficient blast furnace operation, in contrast to the higher iron oxide content of bloomery slag. The addition of lime or feldspar as slag fluidizers lowers slag viscosity, increasing the efficiency of blast furnace operation.

Slater, Samuel: (1768-1835). Builder of America's first automated textile mill on the Blackstone River at Pawtucket, RI, in 1793. After emigrating from England, Slater was able to copy the designs of the machinery invented by Richard Arkwright and used in English textile mills. Unlike England's coke-fired steam-powered textile mills, American industry relied upon water power as its industrial prime mover well into the 19th century.

Slick (slice): **1**) Long chisel for smoothing the planking of a wooden vessel; particularly useful on surfaces not accessible to a lipped adz; also called a slice by many Downeast shipwrights. **2**) Small hand tool used by molders in a foundry to smooth and finish a sand cast prior to the casting process. See core, drag, foundry, hollowware, molder, and sand pattern.

Slip: Grain displacement that occurs within the crystal lattice structure of metals during deformation. See plastic deformation, space lattice, and stress fields, and unit cell.

Slip interference: The collision of crystals within the lattice structure of metals during slip events, which causes internal stress and hardening of the metal. See stress fields and work hardening.

Slip lines: The bands observed on stressed metals due to the deformation of their crystal structure after cold working by any mechanical process. See slip interference.

Slip plane: That component of the lattice structure of a metal that undergoes plastic deformation during mechanical or thermal treatments.

Slitting mill: A mill with machinery used to cut iron bar stock into rods for use in nail making or wire drawing. Often associated with the chafery in integrated ironworks, slitting is a further step in the refining of wrought iron. The first slitting mill was established in the Kent district of England in 1590. Related technology for making wire for wool cards was implemented in England in 1566. Slitting mills were first established in New England in the late 17th century.

Smelt: To reduce a metal ore, e.g. iron, by a chemical reaction in a high temperature furnace; "any metallurgical operation in which the metal sought is separated in a state of fusion from the impurities with which it may be chemically combined or physically mixed" (Shrager 1949, 377). The fundamental chemical process in the smelting of iron ores involves the burning of a carboniferous fuel which reduces the iron oxide ore by removing oxygen, ideally producing pure iron (ferrite) and CO_2. The carbon in the CO_2 is derived from the burning fuel and the O_2 from the reduced ore. Impurities in the iron ore are removed by the addition of flux to the smelting process and are drained out of the smelting furnace as slag. In the smelting of iron ore, the melted or partially melted product is either cast iron produced in the indirect process after cooling/solidification or a semi-pasty bloom of either wrought iron or malleable iron produced by direct process reduction with variable carbon content, usually lower than 0.5%. In the smelting process, the burning fuel removes oxygen from the iron oxide ore, providing iron with a range of carbon content. Varying amounts of iron oxide are not reduced in the smelting process and are transferred to slag as accidental waste products, usually in the form of iron silicate or siliceous slag. See blast furnace, bloomery, furnace types, reduction, siliceous slag, and slag.

Smelting furnace: A furnace used to reduce iron and other metals from their chemically combined state, e.g. iron oxide, to a metallic state by dissolution (reduction) of the chemical bonds of the iron oxides or other iron ores. The most common products of the smelting furnace are wrought iron with < 0.08% carbon content, ranging as low as 0.02% carbon (pure ferrite) and malleable iron with a carbon content ranging from 0.08 – 0.5%. The lower the carbon content of the iron produced by the smelting furnace, the more malleable and ductile is the wrought iron being produced. In the smelting furnace, the iron bloom coalesces at 730° C as a plastic mass that is easy to shape by hammering. If the smelting furnace reaches a temperature of 1150° C to 1200° C, the iron absorbs carbon. At 3.5% carbon content, the eutectic (melting) point is reached, and liquid cast iron is produced, due to the lower melting temperature of iron with a high carbon content. Wertime (1962) notes the difficulty of controlling the smelting process in early shaft and bowl furnaces as follows:

> The small furnaces to be found in ancient China, medieval Europe or modern Africa posses the capability to yield high-carbon cast iron. Any effort to avoid this result and to produce low carbon wrought iron or steel, by means of a process involving only a single step, became a matter of exquisite human and technical restraint. (Wertime 1962, 44)

See bloomsmithing, furnace types (before 1870), natural steel, smelt, and wrought iron.

Smithing forge: Small forge used by blacksmiths, shipsmiths, farriers, cutlers, and other tradesmen to make bolts, horticultural tools, ships' hardware, ox and horse shoes, and hunting knives and swords. The blast was often powered by accordion bellows. Smithing forges were supplied with iron bar stock produced at direct process iron-smelting furnaces or at integrated ironworks from cast iron that had been fined into iron bar stock in fineries and chaferies. After the appearance of the reverbatory furnace and the development of modern rolling mills (1784), wrought and malleable iron bar stock of more uniform quality could be produced for use at a smithing forge. See blacksmith, integrated ironworks, merchant bar stock, and iron-smelting furnace.

Smithy: A blacksmith's workshop for light duty forge welding, in contrast to a larger forge equipped with a trip hammer for heavy duty work. See trip hammer and water-powered drop hammer.

Soaking pits: Modern underground furnaces used to heat steel ingots to a temperature of 2,200° F before rolling and milling.

Softening temperatures: The temperature range at which high carbon steel without alloys softens. The softening temperature range is 600 – 800° F (Pollack 1977).

Solder: A combination of tin and lead used by tinsmiths for joining tin plate since the mid 16th century; now commonly found in contemporary workshops and, recently, the subject of concern due to the health threat from lead poisoning. An alloy of two or more metals used for joining other metals by surface adhesion. See alloy.

Solidification range: The range of temperatures that culminate in the freezing or solidification of a metal. See liquidus and solidus.

Solid solution: The dissolved composition of two metals within each other in their solid state.

Solid solution austenite: Austenite formed when carbon solubility suddenly increases at the critical temperature (Bain 1939).

Solidus: The lower curve in a constitutional diagram, which is the locus of temperatures at which each alloy has completed solidification. See Barraclough's constitutional diagram in *Appendix VI*.

Solingen: An important center in West Germany for the production of cutlery and pattern-welded sword blades, bayonets, and knives; located in close proximity to Cologne and other important steel-producing centers, such as Remscheid and the Rhine River. Solingen was an important steel producing community as early as the Middle Ages; Damascus steel production techniques and samples of Wootz steel may have been brought to Solingen during the Crusades. Thousands of individual toolmakers and swordsmiths used small forges to produce cutlery, swords, and edge tools from the early Middle Ages to the late 19th century.

Sorby, Henry Clifton: The foremost of all English metallurgists; the first scientific researcher to document the microstructure of pearlite and note re-crystallization as the mechanism of recovery of crystalline polarity in ferrous metals (Smith 1960). Sorby, a resident of Sheffield, England, produced the first photomicrograph studies of the structure of iron and steel by polishing and then etching thin sectional slices of various forms of ferrous metals with a dilute solution of nitric acid. He exhibited some of his samples at the British Association in 1863 and wrote a paper on his research about the relationship between the physical properties and microstructure of metals in 1864. However, it was not until he gave a lecture at Firth College in Sheffield in 1882 that his research on the microstructure of ferrous metals became widely known.

Sorinaoshi: The further heating and hammering of the Japanese sword after tempering and the removal of the clay enclosure for the purpose of making fine adjustments.

Source points of biologically significant chemical fallout (BSCF): During the smelting of iron and the production of steel, facilities that burn biomass fuels are source points for biologically significant chemical fallout, including sulfur dioxide and, most especially, methylmercury derived from the burning of coal. Carbon dioxide (CO_2) produced by the combustion of any biomass fuel can also be considered a form of BSCF due to its role in cataclysmic climate change. After 1950, the rapid increases in the production of ecotoxins, including persistent organic pollutants (POPs) derived from pesticide, consumer product, and electronic equipment manufacturing, have dominated the contaminant pulses of BSCF. Nuclear power and fuel and weapons production facilities are source points of biologically significant anthropogenic radioactivity. See biologically significant chemical fallout.

Space Lattice: "The orderly geometric form into which atoms tend to arrange themselves during the process of crystallization" (Shrager 1949). "Every point of a space lattice has identical surroundings... There are 14 space lattices. No more than 14 ways can be found in which points can be arranged in space so that each point has identical surroundings" (Barrett 1943, 2-3). Space lattices are expressed through their system of axes, each of which possesses specific characteristics, i.e. equality of angles and links. The basic unit of a space lattice is the unit

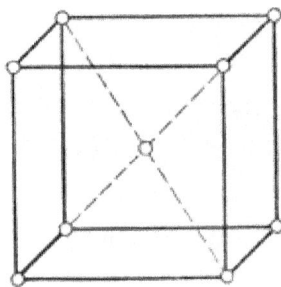

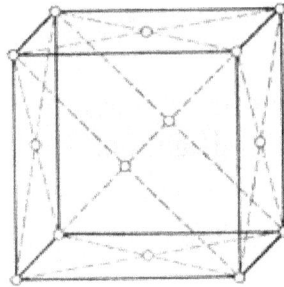

Figure 2. Body-centered cubic unit cell.   Figure 3. Face-centered cubic unit cell.

Figure 1. Space lattice (Shrager 1961, 5)

cell, which, in metallurgy, has three common forms: body-centered cubic (BCC), face-centered cubic (FCC), and close packed hexagonal (CPH). While there are a limited number of forms of the space lattice, there are an unlimited number of crystal structure patterns that can exist within those forms. When in the form of low carbon ferrite, iron is characterized by the body-centered cubic form of the space lattice of its crystal structure, which provides strength in comparison to the face-centered cubic form of ductile metals, such as copper and gold. When heated beyond the temperature of 910° C, the allotropic form, and, thus, the space lattice structure, change to the face-centered cubic form (austenite). Metals that lack plasticity, such as cobalt and titanium, have a close packed hexagonal crystal structure. See allotropic form, Barrett (1943), lattice reorientation, plastic deformation, Shrager (1949), and unit cell.

Spalling: The surface cracking and flaking of metals resulting in the creation of small particles of loose metal.

Spangle: The surface patterns on watered steel swords, derived from the crystal structure of the steel. See Damascus steel, Damascened steel, and especially Cyril Smith (1960).

Spark test: A rule of thumb test often used in the 19th and early 20th century for quickly identifying iron and steel types, also possibly useful in contemporary home workshop situations. The following spark patterns are generated using a powered dry wheel for grinding the item:

Wrought iron
Long yellow streaks, becoming leaf like in shape before expiring

Mild steel
More variety in streak length, with smaller leaves and some sparking

Medium-carbon steel
Almost no leaf, some forking, great variety of streak length, sparking nearer the wheel

High-carbon steel
No leaf, bushy spark pattern, forking and sparking starting very close to the wheel; less bright than medium-carbon steel

Manganese steel
Small leaf before streaks fork to form sparks

High-speed steel
Faint red streaks forking at the tip

Stainless steel
Bright yellow streaks with a small leaf end (Lee 1995, 22)

Spathas: Doubled-edged pattern-welded broad swords used by the Goths to defeat the Romans at the battle of Adrianople (400 AD). In 53 BC, the Romans had defeated the Gauls with the help of their superior swords, made by the same Celtic metallurgists (swordsmiths) in Noricum, who now made the Spathas for the Goths to use against the Romans (Gardner 1892). See Celtic metallurgy, Iron road, pattern-welding, and Styrian steel.

Spathic ore: Another name for the manganese rich iron ore from the Styrian and Carinthian section of Austria. See German steel, manganese, spiegeleisen, and Styrian ore.

Spectrometry: The measurement of spectroscopic interactions. An instrument that performs such measurements is a spectrometer or spectrograph. A plot of the interaction is referred to as a spectrum. The primary form of modern spectrometer used to study ferrous metals is the energy dispersive x-ray spectroscope (EDX), which uses x-rays to generate the electro-magnetic waves that characterize each unique electronic structure. See HRTEM.

Spectroscopy: The study of the interaction between radiation (electromagnetic radiation, or light, as well as particle radiation) and matter.

Spheroidize: To soften high carbon steels by freeing ferrite from cementite, therefore causing the cementite to assume a globular crystal structure, facilitating machinability. Spheroidizing occurs after initial heating just above the transformation temperature, followed by lengthy heating just below the critical point (annealing). See spheroidization and superplasticity.

Spheroidization: Production of round-shaped cementite carbide microstructures, often in veins (nanotubes), by repeated hammering and reheating, for example, samurai and Wootz steel Damascus sword blades. "Forging at temperatures below the temperature of cementite formation from austenite, A_{cm} ($A_{cm} \approx 850°C$) leads to the destruction of needle-like cementite and, after heat treatment, to spheroidization of Fe_3C …Annealing above A_{cm} would made the cementite disappear" leading to the loss of properties characteristic of high quality swords or edge tools, i.e. the combination of toughness (resistance to fracture) and the retention of hardness on the cutting edge (Levin 2005, 905). "Small additions of Mn, Si, P + S seem to enhance the breakup of cementite into spheroidized agglomerates" (Levin 2005, 906). The recent development of the high resolution transmission electron microscope (HRTEM) has allowed a more detailed examination of carbide microstructures in forged steel, resulting in a better understanding of the nanotubes and nanowires that are characteristic of the presence of superplasticity in steels and other metals, including swords and edge tools. See nanotubes, nanowires, spheroidize, and superplasticity.

Spiegeleisen: In its modern form, a pig iron containing 15 - 30% manganese and 4.5 – 5.0% carbon, added to steel both as a deoxidizing agent and to raise the manganese content of the steel (Shrager 1949). Spiegeleisen also refers to cast iron derived from ores containing naturally occurring manganese, which was used to produce "German" steel (1400 to 1875 AD). This cast iron was produced in a blast furnace using manganese-rich siderite iron ores (+/- 2.0% manganese), resulting in cast iron with a bright, mirror-like appearance or "mirror iron," a brittle fracture, and manganese content of 5.0 - 10.0% in its final form. As a component of the flux manganese preferentially combined with sulfur and phosphorus, and their elimination in and as slag allowed more homogeneous carbon distribution within the austenitic iron after the slag formation (manganese oxidation) process. On September 22, 1857, Robert Mushet patented the modern form of spiegeleisen as an addition to Bessemer steel to relieve its "burnt out" quality (Fisher

1963) by enhancing its strength, durability, and hardness. The addition of spiegeleisen was a key step in the successful adaptation of the Bessemer process to a wide variety of industrial applications (e.g. production of steel rails and machinery). The widely varying amounts of manganese in various forms of cast iron produced from 1350 to the present time, and its presence in a wide variety of modern alloy and tool steels illustrates its importance as an alloy of iron and steel. See alloy steel, German steel, manganese, Noric steel, Styrian steel, and tool steel.

Sponge iron: A modern term for the bloom of wrought or malleable iron produced by direct process smelting. See bloom, malleable iron, wrought iron, and direct process.

Spoon adz: Adz used by shipwrights for roughing out large mortises when framing a wooden ship; also called the "gutter adz" in America. Finish work on such mortises was completed with the use of the mortising chisel, lipped adz, and, in some cases, a slick. See adz.

Spring steel: Blister or weld steel that was reheated, re-hammered, and then rolled as sheet saw steel to increase purity and homogeneity. Spring steel was the principal constituent of high quality handsaws before the appearance of cast steel and was also used for coach and other types of springs. Rolled cast saw steel was still labeled "spring steel" on hand saws at least until the late 19th century, as were Henry Disston's "London spring steel" saws. The mechanical and thermal treatments used during these processes served to increase the purity and hardness of the saw steel. Differing mixtures of alloys were sometimes added as micro-constituents during the forging of spring steel. Such formulations were often the subject of extreme secrecy on the part of the steel producers and saw manufacturers, sometimes one and the same. See blister steel, cast steel definition 1, piling, and shear steel.

Sprue: Representative test piece of cast iron, which shows its approximate composition by its fracture. For example, a gray iron fracture indicates a high silicon content.

Spurrier: A smith who forged steel spurs and stirrups in the medieval era (Gardner 1892).

Stainless steel: Steel that contains 18% chromium and 8% nickel, has high tensile strength, and resists abrasion and corrosion, developed by Harry Brearley in 1913 for cutlery. The AISI classification system for steel defines stainless steel as having a minimum of 10% chromium, with or without other alloys. In general, stainless steels are defined as low carbon-high alloy steels. Stainless steels are further grouped according to their heat treatment dependant microstructures: Austenitic – ferritic – martensitic. Martensitic is hardenable by heat treatment; austenitic and ferritic stainless steels are not hardenable by heat treatments. Many types of alloy steels and tool steels contain chromium in lesser quantities. Stainless steel is one of the most important steels produced in the modern electric arc furnace. Martensitic steel (AISI types) 414, 420, 420F, and 440F are often used for the drop-forging of a wide variety of stainless steel hand tools. See the AISI-SAE classification system in *Appendix IV*.

Stamping: The process of manufacturing drop-forged tools or hardware (e.g. gun parts) by forcing red-hot low carbon steel or malleable iron into dies. See drop-forging, low carbon steel, and malleable iron.

Steam-driven tilting band saw: Unique form of power saw for framing out wooden ships. Along with the water-powered circular saw, the appearance of the tilting band saw (c. 1850) was an important step in the mechanization of shipbuilding activities, replacing the pit, frame, and whip saws for many functions.

Steam engine: By 1669, Thomas Savage's steam pump for recovering water from coal mines to save England's threatened forests was improved by the addition of Newcomb's use of pistons to power the pump. Radically improved in the 1760s by Watts' use of improved cylindrical engine pressure vessels, the steam engine became a prime mover of the first Industrial Revolution. In the 1830s, Joseph Nasmyth designed a high pressure steam engine, which was more powerful due to recycled waste steam and played a major role in the development of railroads and, after 1840, the factory system of mass production. In 1884, the development of steam turbines (rotary steam-powered motors) paved the way, after 1900, for the ascendancy of petro-chemical electrical man, and then after 1945, petro-chemical electrical nuclear man where steam engines were the prime movers of nuclear power plants.

Steam turbine: A form of heat engine driven by the kinetic energy of steam in contrast to the piston-driven steam engine, which is powered by the pressure of a volume of steam on a piston. See heat engine and steam engine.

Steel: A thermodynamically stable alloy of iron containing from 0.2% to 2.0% carbon. Steel often contains substantial quantities of manganese ($\pm$1%) and silicon ($\pm$0.6%), and trace amounts of sulfur and phosphorous as microconstituents. In contrast, wrought iron, a silicon slag-bearing iron alloy, is defined as containing <0.8% carbon content. Cast iron, which is much more brittle than steel, is an iron alloy containing more than 2.0 - 2.2% carbon; most cast iron products contain 3 - 3.5% carbon. Note: the upper and lower limits of carbon in steel vary among information sources. A recent source notes that most steels contain levels of carbon from *about* 0.1 - 0.8%, depending upon the desired properties and applications (Sherby 1995, 10-11). The advent of the modern open-hearth electric furnace process allowed the manufacture of a wide variety of steel alloys, therefore opening up the floodgates for tool and machinery production of every description, i.e. the classic period of the Industrial Revolution. The magic of steel is its range of properties. Steel is malleable at high temperatures. Its ductility, malleability, and capacity for being welded decrease in proportion to its rising carbon content. Mechanical and thermal processing dete rmine its microstructure. If containing in excess of 0.5% carbon (the quenching threshold of steel), steel has the unique quality of being greatly hardened by sudden cooling (quenching), after which it is very brittle and fractures easily, requiring further heat treatment by tempering to soften the martinized steel for edge tool production and other uses. The most remarkable quality of steel is its capacity for being tempered to many degrees of hardness, allowing the production of a wide variety of tools and weapons from ancient to modern times. Quality edge tool manufacturing is, in part, contingent on the amount of time spent by the edge toolmaker on the repeated slow reheating, hammering, and slow re-cooling (annealing) of the steel cutting surface of the edge tool. See alloy steel, annealing, bulk process steel, carbon content of iron and steel, German steel, low carbon steel, martenize, microstructure, natural steel, puddled steel, quench, stainless steel, tempered alloy steel, and tool steel.

Steel blooms: Steel produced as semi-solid or solid masses by the following processes: natural, Brescian, German, blister (cementation), and puddling. Liquid steel, in contrast, was produced in crucibles and by modern bulk steelmaking techniques, such as the Bessemer process.

Steel furnace: Prior to the introduction of modern forms of steel furnaces, the term "steel furnace" was synonymous with cementation furnace, which were sandstone furnaces that kept the fire separated from the ore being smelted. Cementation furnaces for blister steel production replaced shaft furnaces for producing German steel in England in the mid to late 17th century and began appearing in America after 1720. See cementation furnace and furnace types.

Steel microstructure: The principal allotropic forms of the microstructure of steel are ferrite, austenite, and cementite. All other variations of the allotropic forms of steel, such as martensite and bainite, are the result of various thermal treatments of steel during quenching, tempering, and annealing. The addition or presence of alloys, such as silicon, manganese, vanadium, molybdenum, etc., may alter the physical characteristics of steel but not their basic microstructural allotropic forms.

Steel piling: Bundling and reforging of raw steel bar stock. The piled steel sword is the characteristic sword of the Vikings, made from piled sheets or strips of steel, possibly Wootz steel obtained during trading activities; piled steel swords are a variant form of the earlier Merovingian swords +/- 700 AD, made from pattern-welded layers of steel and wrought iron. See Merovingian sword, pattern-welding, Viking sword, Japanese samurai sword, and Wootz steel.

Steel press: "A machine for compressing molten steel in casting, to improve the quality of the product" (Audel, 1942, 544).

Steel production: In 1840, the combined steel production of the US, England, and Western Europe was less than 200,000 tons. In 1948, US steel production was about 90 million tons. By 2006, US annual production had dropped to 67 million tons, but world steel production had risen to over 120 million tons *per month*. In the early 21st century, China is the world's largest producer of steel and is building coal-fired methylmercury-producing power plants at a rate of one per week to power its rapidly expanding industrial infrastructure. See biologically significant chemical fallout and mercury.

Steeling: 1) The processes used by a blacksmith to convert red-hot wrought or malleable iron to steel. These processes include repeated hammering to remove slag and other impurities; reheating the iron allowing it to absorb sufficient carbon from a carburizing fuel such as charcoal, for example, by case hardening; quenching (rapid cooling) to achieve hardness; and tempering (slow reheating and cooling) for a final softening of the steel. Prior to the era of modern chemical analysis (1885 f.), a blacksmith judged the quality of his steel intuitively and subjectively, as well as objectively, by considering the following observable properties of his product: appearance, texture, feel, color, fracture, and grain. In essence, the job of blacksmiths was to evaluate the microstructural changes when iron is converted to steel by making visual and tactile evaluations of the products. They observed whether the desired steel product had tears, veins, spots, or fissures indicating impurities or had the smooth look and feel of steel rather than the gray or white grainy quality of cast iron or the fibrous ductile quality of wrought iron. In this context, up until the late 19th century, the work of the blacksmith was an art, not a science. See kan of the ironmonger, carburizing, case hardening, quench, and tempering. **2)** The creation of steel cutting edges on iron tools by scarf welding, or folding and welding steel onto the iron tool, e.g. the creation of a welded steel-iron interface. In the case of knives, steeling was also done by folding iron over steel and then quenching it. Salaman (1975) provides two examples of how hardness required for edge tools is obtained by a working blacksmith: "(a) carbonization: the edge was heated in a charcoal fire and then rapidly quenched in cold water - a method known in pre-historic and Roman Britain, or (b) by the equally old method of 'steeling', i.e. thin strips of good quality steel were let into the edge of a wrought-iron blade and joined by forge-welding in the fire" (Salaman 1975, 247). See ax-making techniques, scarf weld, and weld steel.

Steelmaking c. 1950: Fisher (1949) provides the following concise riff on mid-nineteenth century steelmaking before the era of powdered metallurgy: ore mines, limestone quarries → ore vessels, coke ovens, coal chemical byproducts → blast furnaces → steelmaking furnaces: open-hearth, Bessemer, electric → ingot molds, ingot strippers, ingot soaking pits → semi-finishing mills (blooms, slabs, billets) → finishing mills, rail and structural products; plate, sheet, and strip mills; bar, rod, and wire mills; pipe and tube mills. Fisher then lists approximately four dozen categories of products produced by the finishing mills ranging from rails, bridges, automobiles, ships and aircraft to tools, springs, farm machinery, and other consumer products (Fisher 1949, 60-61).

Steel scale: The flaking layer of iron oxide produced on steel surfaces as a result of oxidation, also called fire scale. See iron oxide.

Steelsmith: A bloomsmith whose only function was to produce raw steel rather than wrought iron. In the early Iron Age and prior to the evolution of the blast furnace and the indirect process of producing wrought iron and steel, most steelsmiths would have been associated with furnaces producing raw steel for sword makers. Contemporary Japanese steelsmiths continue to smelt raw steel for samurai swordmakers in breakdown furnaces, as depicted in a *Nova* episode on PBS. See bloom, bloomsmith, breakdown furnace, natural steel, and pattern-welding.

Stock removal method: A modern knife-making technique involving sawing and then grinding off excess steel from steel bar stock, in contrast to the forging of the knife blade (Latham 1973).

Stopped off: The covering over of parts of an iron tool, other than its working edge, during the carburization of edge tools that occurs in forge welding. Clay was a common media for enclosure during such procedures; the clay cover protected the iron from the oxidizing impact of burning fuel. See carburize, enclosure, and forging.

Stour Valley: An important 18th century center of furnaces and forges in the Severn River watershed in the Midlands of England. The ironworks in the Stour Valley supplied many of the toolmakers of nearby Birmingham in the years before Sheffield became England's most important steelmaking center.

Stourbridge: An early steelmaking center in the Midlands of England, famous for its fine heat-resisting clay, so important to Benjamin Huntsman's production of crucible cast steel (1742). See cast steel, crucible, and Stourbridge clay.

Stourbridge clay: The high temperature resistant clays found near the Village of Stourbridge in England and utilized by Benjamin Huntsman to make the crucibles used for the production of the cast steel that he needed for his watch spring business. No such high temperature resistant clays were available in the U.S. for production of cast steel until the 1840s, when Joseph Dixon developed the graphite crucible. See cast steel definition 1 and crucible.

Strain: Stress within the microstructure of ferrous metals caused by grain boundary sliding and grain rotation.

Strain hardening: The deformation of the crystals in the lattice structure of metals, which increases their hardness and yield strength prior to shear stress-induced slip. Strain hardening decreases with rising temperature. Annealing is the recovery from the strain hardening of the lattice structure of steel, for example, which is then softened and made less brittle (under less strain) by slow temperature changes in the process. See annealing and hardenability.

Strain rate sensitivities: Sensitivities that increase with increasing temperature and with decreasing grain size.

Stress: Internally distributed forces within the microstructure of metals in the form of the mechanical reaction of the microstructure to deformation forces. "Stresses always occur in pairs" and include compressive, tensile, and tangential stresses (shearing) (Baumeister 1958, 5-23).

Stress fields: Fields that surround dislocations on slip planes. "Dislocations may be generated at a flaw and follow each other along a slip plane until they encounter some imperfection or a boundary where they become stuck. The stress field that they set up at this point opposes the approach of additional dislocations and therefore hardens the metal" (Barrett 1943, 338). Stress fields may also build up a super lattice of dislocations, which serve to raise critical shearing stress through the phenomena by which similar dislocations repel each other and dissimilar dislocations attract each other, creating a checkerboard of stress fields throughout the crystal. See creep, plastic deformation, martensite, slip, shearing movements, and work hardening.

Stress relief: Thermal treatments, such as tempering, that relieve the stresses in the lattice structure of iron and steel alloys caused by cold working, quenching, and work-related plastic deformation. See annealing, cleavage plane, creep, and space lattice.

Structural unemployment: A consequence of the revolutionary impact of the Age of Information Technology. A vast underclass of workers is confined to low paying service jobs instead of higher paying and now computerized manufacturing (etc.) positions. Structural unemployment provides the context for future viability of an underground economy of skilled craftspersons and artisans essential for the maintenance of the global market economy infrastructure. Knowledge of the history of technology and an ability to use hand tools will be key components of the viability of alternative economies, which are not characterized by structural unemployment.

Stucköfen furnace: The high shaft furnace used in Austria and Germany during the Renaissance that evolved out of earlier low shaft furnaces and was the immediate predecessor of modern blast furnace forms. See furnace types.

Styrian ore: Iron ore with 1.0 - 2.0% manganese content found in the Styrian-Carinthian area of Austria but otherwise not commonly encountered in the ores of Sweden and the United States. This ore facilitated the direct process production of iron having steel-like qualities (strength, durability, hardness), while still containing significant amounts of siliceous slag inclusions. This ore also played a major role in the production of German steel. The high manganese content of cast iron assisted in the removal of the unwanted contaminants sulfur and phosphorous in the flux. See German steel, natural steel, manganese, spiegeleisen, Styrian steel, Weald of Sussex, and Weardale.

Styrian process: The continental technique of refining (decarburizing) pig iron containing manganese (spiegeleisen) by the use of charcoal fuel to produce wrought iron, malleable iron, or German steel.

Styrian steel: Natural steel produced by the reduction of iron ores containing manganese in direct process bloomeries, later produced by fining (decarburizing) blast-furnace-derived spiegeleisen; also called German steel. The Styrian method, also called the continental method, was the principal steel-producing process of Renaissance England and Europe north of the Alps between 1350 and 1650. Carinthia and Styria, in Austria, were the principal sources of manganese-containing iron ores from the early Iron Age until the 19th century, thus the name of the process. Before 1600, manganese-laced ores were also found in the Weald of Sussex (England). See continental method, German steel, natural steel, and Noric steel.

Sulfur: Nonmetallic chemical element (S) that is a naturally occurring microconstituent of most iron ores and a major contaminant in iron and steel production resulting from the introduction of coke as a fuel. Sulfur limited the uses for cast iron and steel unless neutralized by a high lime base slag at high temperatures, producing the phenomena of "hot shortness" during forging, when steel contaminated with sulfur would disintegrate upon impact when hammering. The introduction of steam-driven piston blowers, creating higher temperatures that helped burn out the sulfur, allowed the successful use of coke as a universal blast furnace fuel in the late 18th century (Tylecote 1976). Sulfur is also neutralized in slag at lower melting temperatures by the presence of manganese in iron ore in the smelting process. See hot short, manganese, and spathic ores.

Superplasticity: "Polycrystalline solids which have the ability to undergo large uniform strains prior to failure" (Sherby 1995). Superplasticity is further defined as the capacity of a metal or other polycrystalline solids in tension at high temperatures (above ½ of the melting temperature or eutectic) to undergo deformation elongations above 200% prior to failure (mse.mtu.edu). "The explanation of the superplasticity of the UHCS [ultra high carbon steels] is that the typical microstructure can lead to a fine uniform distribution of spheroidised cementite particles (0.1 mm diameter) in a fine grained lower carbon ferrite matrix" (Srinivasan n.d., 59). In his classic *Functions of the Alloying Elements in Steel*, Bain (1939) notes superplasticity results when "coarse networks of pro-eutectoid cementite forms along grain boundaries of prior austenite producing a uniform distribution of fine spheroidised cementite particles (.1 μ diameter) in a fine grained ferrite matrix." In ferrous metallurgy, the most notable historic example of a steel having characteristics of superplasticity at high temperatures is the Damascus steel sword, famed for its ability to bend while retaining a hard cutting edge. The probable source of the superplasticity of such swords is cementite microstructures in the form of nanotubes (spheroids) created by the forging of very high carbon steel (1.5 - 2.0% cc) that contains significant micro-contaminants such as cobalt, vanadium, or manganese. Forging at temperatures well below 850° C results in spheroidization of the needle-like structures of cementite, creating cementite nanotubes with their characteristic nanowires. High temperature superplasticity facilitates plasticity and toughness at normal temperatures in Damascus and samurai swords, as well as in forge welded edge tools. The molding capability of superplastic metals, i.e. their ability to exactly replicate the forms of CAD (computer assisted design) industrial equipment patterns, makes them the most important component of modern machinery production, e.g. aircraft engine turbines. The role of alloys, such as vanadium and cobalt, as microconstituents of the cementite nanowires in facilitating superplasticity has not yet been determined. See CAD, carbide forming metals, cementite plates, Damascus sword, electron microscope, HRTEM, nanotubes, nanowires, and patternmaker.

Surface carburizing: A difficult, tedious, and infrequently used method of steel production prior to the era of modern steel production that required protecting the tool being carburized from the combustion gasses, which would oxidize (decarburize) the iron or steel surfaces of the tool being forged. In the case of the cementation process used to produce steel (<1830), surface carburization was the first step in a process that would continue for days or weeks. See carburize, case carburization, case hardening, cementation process, enclosure, and nest carburization.

Surface decarburization: The propensity of the surface of some carbon steels to lose carbon if heated above 1333° F. The hardness of such surfaces is drastically reduced when quenched from elevated temperatures (Pollack 1977). See carburize and edge tool forging.

Surface hardening: To impart to the surface of low carbon steel, by carburizing (austenizing followed by slow cooling), a wear and abrasive resistant surface that retains a ductile and tough interior composition; also called case hardening. This is a modern steelmaking technique. See carburize, cementation furnace, and case hardening.

Surface oxidation: Result of exposure to oxygen in the atmosphere, which causes the oxidization of carbon and scaling on the surface of the iron tool being worked. The higher the forging temperature, the higher the rate of surface oxidation is. Surface oxidation is the most significant challenge in the successful hot forging of iron implements or tools. See oxidization.

Swage: Tool used by blacksmiths for rounding bar stock. The swage has two components. The bottom swage fits into the hole in an anvil or swage block. Top swages are often handled and used to pound the metal into the desired shape on the bottom swage.

Swage block: A large block of cast iron, semi-steel, or wrought steel with multiple holes and rectangles of various sizes used for shaping hot bar stock by a blacksmith or toolmaker. See dapping block.

Swedish iron: The high quality, low sulfur, nearly phosphorus-free, highly refined wrought iron derived from charcoal-smelted cast iron and exported from Sweden to England and America. English steelmakers preferred Swedish charcoal-smelted iron to domestically produced mineral-fuel-smelted iron because of its superiority for blister and crucible steel production. The success of the English steel industry (1750-1860) was closely connected to the ready availability of Swedish iron. Despite international trade restrictions, Swedish bar iron was imported to the colonies and to toolmaking centers, such as New Bedford, in the early 19th century, possibly for use in steel furnaces for blister steel and whaling tool production. See blister steel, charcoal, charcoal iron, finery, Whalecrafters, and wrought iron.

TEM: Transmission electron microscope (TEM), which utilizes a high voltage electron beam emitted by a tungsten filament cathode to produce an image of the inner structure of the metal being studied, which is then magnified by a series of electro-magnetic lenses and recorded on photographic plates, fluorescent screens, or light sensitive cameras. See electron microscope, HRTEM, REM, and SEM.

TRIP: See transformation induced plasticity.

Tamahagane: The Japanese steel used for the production of samurai swords. It is produced from locally available iron sand known as "satetsu," which is smelted into raw steel in the Japanese open-hearth clay breakdown furnaces known as "tataras." See Chalybean and Japanese swords.

Tap hole: The discharge opening for molten metals in a smelting furnace.

Tatara: The Japanese term for the furnace used by samurai sword makers for smelting raw steel out of iron sand (satetsu). A form of breakdown furnace, the tatara is an open-hearth clay furnace approximately four feet by twelve feet and is deconstructed after every firing for the removal of the raw steel bloom. See breakdown furnace.

Taylor-White: Frederick W. Taylor and Maunsel White developed the modern improved form of high speed steel by heating high quality alloy tool steels to a temperature just below their melting temperature (1900). When these steels cooled, they were able to maintain efficient cutting capabilities at much higher temperatures and much greater speed than the older version of alloy steel, R. M. S. Widely used after 1910, the use of Taylor-White steels required heavier machinery and, thus, the retooling of the many industries who used high speed steels in the 20th century. Molybdenum became an important addition to tungsten in alloy tool steel configurations. Taylor and White also discovered, during their investigation of high speed steels, that a key ingredient along with tungsten in Robert Mushet's self-hardening steel (R. M. S.) was its 2.0 – 2.5% manganese content (Tweedale 1986). See manganese, Mushet, Robert F., and R. M. S. steel.

Tea Party Taliban: America's answer to the rise of radical jihadists. Their highly partisan radicalism is a reflection of the growing social stress characteristic of the unfolding Age of Biocatastrophe and one of many challenges to the viability of sustainable economies of the future. The Tea Party Taliban joins greed and hyper-digital technology as the primary enablers of a klepto-plutocracy based on institutionalized income disparities.

Techno-elite: Born-again ironmongers in a world of silicon chips. Where did the manganese go? The techno-elite, skilled born-again ironmongers, are the primary enablers of the efficient functioning of the complex infrastructure of modern global consumer society. The success of modern information and communications technologies is dependent upon their skills and ingenuity.

Teeming: To use a ladle to fill ingots with molten metal.

Temp: The trough, usually made of stone, that transported melted iron from the blast furnace to the molds used for making hollowware. See blast furnace.

Temper: To reheat steel that has been hardened by quenching (martensite) to reduce the hardness and especially the brittleness of edge and other tools, usually done at a temperature range of 150° C to 650° C for periods ranging from 30 minutes to hours (Barraclough 1984a); closely related to annealing. The variables of time and temperature determine the microstructure of steel, as with martensite, where tempering precipitates iron carbide (cementite), creating less brittle steel by encouraging the homogenization of its carbon content. "To make steel less brittle and therefore tougher, to relieve internal stress and either to stabilize the retained austenite or to cause it to transform into a dimensionally stable structure" (Shrager 1949, 165). See austenite and martensite.

Temper carbon: The form of carbon found in grey cast iron and malleable cast iron in the form of nodules of large gray flakes of free graphite; also called free carbon. Temper or "free carbon" softens cast iron, weakening its structure, but also increasing its machinability. Annealing of cast iron can restore durability, especially in cast iron containing silicon, manganese, and other alloy micro-constituents. The annealing of grey cast iron and its temper carbon content were key components of the successful production of iron-bodied hand planes by Leonard Bailey & Co. and later Stanley Rule and Level Co. (1850 to 1870f.). See combined carbon, free graphite, grey cast iron, malleable cast iron, and white cast iron.

Temperature control: The key to successful forging techniques for the earliest edge tool and edged weapon makers, as well as for the most modern knife, edge tool, and tool steel manufacturers. Loss of accurate temperature control during the forging process results in an inferior product. The same observation can be made about hardening alloy steels or austenizing and annealing any form of steel or cast iron.

Temperature range for melting iron and steel: Pure wrought iron (cc < 0.03%) melts at a temperature of 2793° F / 1538° C. Increasing carbon content results in the lowering of the melting temperature such that cast iron (cc > 3.5%) has a melting temperature of 1149° C.

Tempered alloy steel: A now obsolete term that came into widespread usage after Robert Mushet invented his "self hard" tungsten-manganese alloy steel in 1868. Toolmakers after this date combined new discoveries about the efficacy of tempering and annealing techniques with the adaptation of alloy steel in hand tool production, especially for machinists' tools. After Taylor and White developed a wide variety of new tool steels and Harry Brearley perfected the production of stainless steels in 1913, production of tempered alloy steels became so widespread that this designation disappeared from tool markings just as the designation "cast steel" was no longer used to mark edge tools. See alloy steel, drop-forging, Mushet, Robert F., stainless steel, and Taylor-White.

Tempered martensite: In fully tempered steels, the change from the martensite matrix to homogenous distribution of carbon is a result of the uniform distribution of fine particles of cementite within the martensite, stabilizing its lattice structure by relieving its brittleness (i.e. its strain hardening). Acicular carbide units of martensite grow in size, decrease in number, and become more spherical, increasing plasticity. See austenite, cementite, critical point, superplasticity, and temper.

Tensile elongation: The tensile elongation capacity of steel at room temperatures decreases as carbon content increases. Low carbon steel (< 0.05% cc) shows significantly larger percentages of elongation, up to 30% on an 8" test specimen than higher carbon

154

steels, which approach 0% tensile elongations in steels with a carbon content of 1.3%. Crucible steel has the highest elongation plasticity among test steels (Howe 1890).

Tensile strength: A measurement of the maximum load that a material can withstand without being pulled apart; also known as maximum strength or ultimate strength. Tensile strength is measured in units of force per unit area. The non-metric units are pounds-force per square inch (lbf/in^2 or psi). Soft grey cast iron has the lowest tensile strength at 23,000 lbf/in^2 or psi. Tensile strength is highest in high carbon tool steels used for lathe tools, chisels, files, and saws (steel with a 1.25% carbon content has a tensile strength of 135,000 psi); wrought iron has an intermediate tensile strength of 52,000 psi.

Tetragonality of martensite: The formation of the anomalous atomic arrangement of the microstructure of martensite due to the failure of carbon atoms to form carbides at lower temperatures. As a result, the formation of asymmetrical, tetragonal, ferrite-cementite acicular, plate-like microstructures gives hardness and brittleness to low carbon martensite. Martensite can be uniformly softened by reheating (annealing), which modifies its acicular plate-like microstructures. See bainite, lattice structure, and microstructure.

Thermal fluctuations: Changes in temperature that affect the stress fields of the crystal lattice structures of metals and that thus generate lattice dislocation. See annealing, lattice reorientation, and tempering.

Thermal hysteresis: The lag time during heat treatment, characteristic of either cooling (*r*) [refroidissement] or heating (*c*) [chauffage]. Minor variations in heating and cooling rates can play a major role in the phase transformation kinetics of alloy steel production, especially for hardened steels used in applications where superplasticity is essential. See critical temperature range and transformation temperature.

Thermal stress: The stresses in the crystal lattice structure of metals caused by temperature change. See thermal fluctuations.

Thermal treatment: The heating and cooling of iron and steel, which produces a range of structures from coarse pearlite to fine martensite. See austenize and martensite.

Thermit welding: Fusion welding utilizing finely divided aluminum and iron oxide (thermit), principally used for welding rail joints. A mold is constructed on either side of the rails to be welded and the rails are then superheated (2800° F), which ignites the thermit. The reaction spreads throughout the mass. The aluminum unites with the oxygen molecules of iron oxide producing pure iron in the form of superheated (up to 5000° F) aluminum alloy liquid steel. Discovered in 1896, thermit is produced in refractory lined crucibles. The process required preheating the rails. Thermit welding is no longer used for rail welding.

Thermophysical properties: Differences in the allotropic phases of iron and steel due to the absorption or release of heat and consequent volume changes due to microstructural deformations (shearing, diffusion, etc.)

Tilt hammer: A variation in the form of the trip (helve) hammer. The tilt hammer first appeared as illustrations on Greek vases depicting early ironworks. It operates with a gear mechanism and fulcrum, differing slightly in design from the trip hammer, which has a revolving fulcrum mechanism (Horsley 1978). Modern lightweight tilt hammers were designed in the mid-19th century, which operated at higher speeds and facilitated the mass production of a wide variety of tools and machinery. See helve hammer.

Tin plate: Sheet steel covered with a thin layer of molten tin. See whitesmith.

Topman: In the operation of the blast furnace, the person feeding the furnace with ore, charcoal, and flux.

Tool: Any instrument of manual operation (Oxford English Dictionary 1975).

Tool and die steel: Another name for tool steel and high speed tool steel. See tool steel.

Tool placement awareness (TPA): Easily learned applied mental discipline that has been demonstrated to significantly reduce tool misplacement (misplace: to stow with no hope of retrieval). A bar graph describing the process of misplacement shows the tool interest curve rising rapidly during a search, sometimes with a parallel rise in blood pressure, thence to level off during the period of useful possession, only to drop abruptly when the task is done and the tool is cast aside mindlessly. A fractionally different approach persuades the user to maintain the high tool interest plateau for a moment past the period of useful possession, perhaps in that time giving thanks for the privilege of tool access and only then releasing the tool mindfully to a place where it can be retrieved efficiently. The key to either frustrating misplacement or efficient retrieval is held in that moment when hand and tool separate. Think about it, and, when necessity or whim demands, the reunion will be swift (McLaughlin, David 2008, Liberty Salvage Co., personal communication).

Tool steel: 1) In general, any steel with a carbon content at or above 0.5%, which can then be effectively quenched and tempered to create a more homogenized carbon content. Modern tool steels used for machinists' cutting tools are defined as having a carbon content range of 0.7 – 1.5%. Tool steel is also defined as steel with a carbon content of 0.5 – 1.3%; used for the working parts of a tool (Palmer 1937). Other sources define tool steel as having up to a 2.0% carbon content. Tool steels may also contain other alloys, including silicon, manganese, cobalt, vanadium, chromium, and molybdenum. See classification system for steel, alloy steel, and carbon content of iron and steel. 2) Modern tool steels are now differentiated from standard steels (see carbon steel classifications) and are now classified as any steel containing more than 4% of one or more alloys, which increase their ability to cut steel at high speed, hence the name "high speed tool steel." Modern tool steels may contain tungsten, which increases the resistance to the softening affect of temperature and forms abrasion resistant carbides. Vanadium will increase forge-ability while also increasing hardness, wear properties, and carbide content. Molybdenum increases deep hardening and is now considered a cost effective substitute for tungsten. Chromium increases hardenability and corrosion resistance, and in

156

conjunction with higher carbon content also increases toughness and wear resistance. Cobalt high speed tool bits are especially useful for drilling heat resistant alloys and materials with a Rockwell hardness exceeding 38c. Multiple and diverse combinations of alloy elements are considered to have a synergistic impact on the efficiency and durability of the several hundred modern variations of high speed tool steels used as cutting tools in CNC milling machines and other equipment. See carbon content of steel, classification of tool steels in *Appendix V*, Mushet, Robert F., and Taylor-White.

Tough iron: A slang term for the high quality, low phosphorus iron produced by the bloomery forges of the Forest of Dean in the decades preceding the English Civil War (1645 – 1653); also used to refer to any high quality low phosphorus wrought iron.

Toughness: The ability of a metal to withstand impact without fracture; the opposite of brittleness, commonly used to express resistance to sudden shock. Due to its entrained slag, wrought iron is famous for its tensile strength and toughness. See mechanical properties of metals.

Tragedy of the round world commons: The unfortunate impact of four centuries of cascading Industrial Revolutions and a now globalized consumer society on the vulnerable, nonrenewable finite resources of our biosphere. The impact of innovative flat world information and communications technologies may exacerbate the tragedy of the round world commons by diverting our attention from a biosphere in crisis.

Transcrystalline fracture: Fracture that passes through the lattice structure of metals rather than around the boundaries of crystal structures (intercrystalline fracture); "the normal type of failure observed in metals" (Shrager 1949, 380).

Transformation: With respect to ferrous metallurgy, changes in the allotropic form of steel, also called phase change. Transformation mechanisms are either homogenous, characterized by uniform distribution, as in the fine carbide spheroid microstructures of cast steel, or by heterogeneous distribution of ferrite-pearlite microstructures in raw steel. See allotropic forms of iron, microstructure, and transformation temperature.

Transformation induced plasticity (TRIP): Modern term for the study of the characteristics of bainite reactions, the mechanisms of which are not yet fully understood (Sinha 2003).

Transformation temperature: Temperature levels at which phase change occurs in iron carbon alloys.

> There are three transformation temperatures, often referred to as critical temperatures, which are of interest in the heat treatment of steels.
>
> - [Eutectoid temperature: $723°$ C (A_1)] The boundary between the ferrite-cementite field and the austenite-ferrite or austenite-cementite field.
>
> - [Alpha-gamma iron transformation temperature for pure iron: $910°$ C (A_3)] The A_3 line represents the boundary between the ferrite-austenite and austenite fields.
>
> - [Critical temperature range: The temperature difference between (A_1) and (A_3) expressed

as (A_{cm}).] The A_{cm} line is the boundary between the cementite-austenite and austenite fields.

The thermal hysteresis [lag] that occurs on heating is indicated by the letter c, representing the French word *chauffage*, meaning heating. Similarly, thermal hysteresis on cooling is indicated by the letter r, representing the French word *refroidissement*, meaning cooling. Thus there are two sets of critical temperatures: Ac_1, Ac_3, and Ac_{cm} for heating and Ar_1, Ar_3, and Ar_{cm} for cooling... The faster the rate of heating, the higher the Ac point: the faster the rate of cooling, the lower the Ar point. Thus the faster the heating and cooling rates, the larger the difference between the Ac and Ar points of the reversible equilibrium point A (Sinha 2003, 1.7).

See austenized, critical temperature, and the iron carbon diagrams in *Appendix VI*.

Transformation time: The amount of time needed to produce various microstructures by quenching or tempering austenite. Pure martensite is produced by rapid quenching (cooling) of austenite to a temperature of 220° C in 80 seconds or less, while the actual transformation occurs in less than one second. Martensite-bainite complexes are formed with cooling rates of 80 to 2,000 seconds, martensite-ferrite-bainite microstructures have a transformation time range of 2,000 to 30,000 seconds, and martensite-ferrite-pearlite-bainite microstructures have a transformation time range of 30,000 to 100,000 seconds (69 hours). Very slow cooling, as in the production of cementation steel, produces ferrite-pearlite microstructures. See the transformation time diagram from Pollack (1977) in *Appendix VI*.

Transition point: Temperature at which changes occur in the allotropic forms of metals.

Transmission electron microscope: See TEM.

Treenail: Wooden pegs, pronounced and often spelled "trunnel," usually 8 to 14 inches long, used to secure the framing of a ship. Wooden treenails were superior to iron nails and spikes because they didn't rust over long periods of time and swelled when exposed to moisture after ship launching, thus tightly securing the ship frame constituents.

Trip hammer: A water-powered hammer developed during the medieval period; also called a helve hammer or drop hammer. The widespread appearance water-powered trip hammers coincided with development of the blast furnace after 1350 and represented the first step in more efficient iron production by relieving the bloomsmith of much of the work of hammering out the slag from the bloom. Operating initially at 50 – 75 blows per minute, trip (helve) hammers increased in size and speed during the 18th and early 19th centuries. Joseph Nasmyth invented the steam-powered trip hammer in 1838. In America, coal- or coke-fired steam hammers made it possible for forges to be located well away from water power sources, encouraging centralized development of iron production centers. In the 19th century, large trip hammers as well as smaller tilt hammers played a central role in the factory system of the mass production of drop-forged low carbon steel tools. The increasingly powerful hydraulic press has replaced most trip hammers since 1875. See drop-forging, drop hammer, helve hammer, overshot water wheel, shingling, and tilt hammer.

Trompe: Hydraulic bellows in the form of a water-blowing engine run by air compressed by falling water. Pipes carry air and water to a receiver where the air is forced into a blast pipe connected to the tuyère, providing the air for the smelting process. See blast furnace and tuyère.

Trunnel: See treenail.

Trunnel auger: A wormed auger used to cut the holes for wooden trunnels (treenails) to secure the framing of a ship (in lieu of spikes and bolts). Characterized by double spiraled flutes, two cutting edges, and two lips, trunnel augers were a common product of a shipsmith's forge before the era of the factory system of tool production.

Trydimite: SiO_2, noted as a microconstituent in Damascus Wootz steel swords subject to x-ray phase analysis (Levin 2005).

Tsuba: The hand guard of a Japanese sword.

Tsuchioki: Application of mixtures of adhesive clay to the blades of Japanese swords prior to preferential tempering. The thickest layer of clay would be applied to the top and upper sides of the sword, with a much thinner layer applied to the steel cutting edge. The uniform heating of the blade above the critical temperature resulted in more rapid cooling of the cutting edge, which, due to preferential tempering, was sharper (because it contained more martensite) than the softer sword body. See critical temperature, enclosure, samurai sword, temper, and yaki-modoshi.

Tungsten: Important alloy in alloy and tool steels. If present in amounts of +/- 4%, it imparts hardness and wear resistance. When present in amounts of about 18%, it imparts red hardness and hot strength to steel. It forms abrasion resistant particles in tool steel, promotes carbide formation, and inhibits softening during tempering by facilitating secondary hardening (Pollack 1977). See the AISI-SAE system of designating carbon and alloy steels in *Appendix IV*, alloy steel, Mushet, Robert F., and tool steel.

Tungsten steel: Alloy containing 5 to 7% tungsten. It is used for manufacturing magnets and in tool steel to facilitate cutting (milling).

Turbine: A machine in the form of a rotary engine that extracts energy from a fluid flow. The first turbines, as prime movers, were the windmill (an air turbine) and the water wheel (a water turbine). These were followed by the development of the steam turbine, gas turbine, and jet engine. See machine and simple machine.

Tuyere (tuyère): The nozzle at the base of any furnace or forge for the admission of the air produced by a leather bellows or any other air-blast-producing device. The forced air draught (air blast) needed for the smelting process would enter this nozzle and follow the tuyère tunnel to the charcoal fuel being burned. The use of the tuyère was the first of the improvements in furnace design and iron-smelting of the early Iron Age. See blowing devices.

Twinning: Reorientation of lattice structures during plastic deformation of metals. Along with slip, it is one of the principal forms of crystal deformation. Twinning is one of the results that occur when "a lamella within a crystal takes up a new orientation related to the rest of the crystal in a definite symmetrical fashion. The lattice within the twinned portion is a mirror image of the rest; the plane of symmetry relating one portion to the other is called the twinning plane" (Barrett 1943, 307). In many cases, twinning involves rapid slip or sheer along the glide plane, rather than rotation of the planes, and is the characteristic deformation process that occurs during annealing. Barrett also defines twinning as the symmetrical reorientation of exactly duplicate lattice structures occurring rapidly during slip events by the "…shearing movements of the atomic planes over one another" (Barrett 1943, 308). "…twinning brings slip planes into position for further deformation" (Barrett 1943, 316). See deformation twinning, lattice structure, and Neumann bands.

Ulfberht: Signature of a medieval swordmaker or a family of swordmakers noted on over 100 swords found in a wide ranging number of locations in continental Europe (Williams 2007).

Ultra fine microstructures: The fundamental characteristic of ultra high carbon steel (UHCS); "ultra fine spherical particles of iron carbide embedded in ultra fine spherical grains of iron" (Sherby 1995, 12).

Ultra high carbon steel (UHCS): Steel with a carbon content of 1.2 - 2.0%. Only ultra high carbon content steels have the capacity to exhibit characteristics of superplasticity at high temperatures above ½ of their eutectic.

Ultra low carbon bainitic (ULCB) steel: A modern form of heat-treated steel with a very low carbon content (0.01 - 0.03%) characterized by toughness and strength; especially used for piping in Arctic climates (Sinha 2003, 9.31). This steel has the same carbon content and physical characteristics as wrought iron but has no ferrosilicate in its chemical composition.

Undercool: To cool quenched austenite below its critical temperature but not so low as to reach the critical temperature needed to form martensite (220° F), i.e. the arresting of the cooling process; a strategy used to produce steel with bainite microstructures; particularly useful in the production of hardened tool steels. The generic definition of undercooling is "the lowering of the temperature of a liquid beyond the freezing temperature and still maintaining a liquid form" This process results in homogeneous nucleation and is also known as super cooling (Tom Rahtz/UAH, http://science.nasa.gov/ssl/msad/dtf/under1.htm). See austenite, bainite, and martensite.

Underground economy: An already flourishing invisible network of artisans, artists, organic farmers including jewelers, silversmiths, weavers, and born-again ironmongers of every description. An inevitable social and economic response to the growing economic power of predatory corporations, both national and multinational, and the gross income and educational inequality they foster. See klepto-plutocracy.

Undershot water wheel: Water comes underneath the wheel, driving it in a clockwise direction in contrast to the counterclockwise direction of the overshot water wheel. Undershot is the oldest form of water wheel. See overshot water wheel.

Unified numbering system: System developed and published by the Society of Automotive Engineers, Inc. and the American Society for Testing and Materials (1986), which lists over 3,000 identifier designations for metals and alloys. See *Appendix III*.

Uniform transformation: The production of high alloy steels with minimal internal stresses by slow transformation processes, such as normalizing in still air or by cooling rates, which are slower than those needed to produce martensite. "Oil quenching produces less rapid change of volume and consequently less distortion than does water quenching" (Shrager 1949, 157), hence its preferred use in the edge tool forging process. See normalizing and quench.

Unit cell: Basic unit of the space lattice of the crystal structure of metals that expresses their allotropic forms. Iron has two forms of temperature-dependent unit cells: the body-centered cubic structure (BCC) up to 910° C (1670° F) (alpha iron,) and the face-centered cubic structure (FCC) from 1670° F (910° C) to 2552° F (1400° C) (gamma iron). Above 1552° F, iron reverts to the FCC unit cell structure. A third form of the unit cell is the close-packed-hexagonal (CPH) space lattice structure, characteristic of non-ferrous metals, such as cobalt, magnesium and titanium. See allotropic forms, body-centered cubic structure, close-packed hexagonal structure, and space lattice.

Unit cell parameters: The multiplicity of the patterns of crystal microstructures for example, in cementite in Wootz steel swords, which may result from the substitution of the carbide-forming alloys V, Mo, Cr, Mn, and Co for Fe atoms, or Si, S, and B for carbon atoms within Fe_3C structures (Levin 2005).

Upset: To make a piece of iron rod thicker and shorter by hammering. See Bealer (1976).

Vanadium: A rare chemical element (V) occurring in iron, uranium, and lead ores. Vanadium is a deoxidizer that greatly increases hardenability, is a strong carbide former, elevates grain coarsening temperature (thus yielding fine grained steel), and resists tempering by aiding secondary hardening (Pollack 1977). An important tool steel alloy, vanadium also increases the strength of pearlite by precipitation strengthening of ferrite and increases the ductility of hypereutectoid steels (Levin 2005). As an important alloy, it is a very scarce and, thus, expensive metal. See alloys and hardenability.

Venetian steel: Steel produced or available in Venice in the 17th century; noted by Moxon in 1677 as the highest quality of all the steels available in the 17th century. It is not currently known what techniques or combinations of techniques the Venetians used to produce their famed steel. Natural steel from Noricum had been brought down through Alpine passes to Italy during both the Roman Republic and Roman Empire. Brescian steel was ubiquitous during the Italian Renaissance, and Wootz steel, the smelting techniques of which were closely related to those used to produce Brescian steel, would have been readily accessible to Venetian ironmongers. See Brescian steel, Moxon, and

Wootz steel.

VG-10 stainless steel: A high quality grade of stainless made in Japan for use by food and sports cutlers. It is composed of carbon: 1.0%, chromium: 15.0%, molybdenum: 1.0%, vanadium: 0.2%, and cobalt: 1.5% (Erin Casson 2013, personal communication).

Viking sword: Pattern-welded all steel sword blade similar in its layered forging to the iron and steel pattern-welded Merovingian blades of central Europe (+/-750 AD). The earliest sword of this type was found in a Roman site in Britain in the 2nd century. It was the principal weapon of the Vikings and probably locally made. Smith (1960) notes their sudden disappearance in the 10th century. It is unknown whether the well-traveled Vikings obtained some of the metalworking skills from Muslim countries making Wootz steel (Smith 1960).

Walloon process: A process that uses charcoal-fired, open-hearth forges to fine cast iron into wrought iron from blast furnace pig iron, commonly employed starting in the late 14th century to 1784, when Henry Cort perfected the puddling furnace. Charcoal protected the bloom of iron from being oxidized prior to its removal for forging. The Walloon process furnace was used to produce Swedish iron, as well as to fine the pig iron at the Saugus, MA Ironworks c. 1646 (Gordon 1996). The Walloon furnace produced not only very low carbon wrought iron (0.02 - 0.08% carbon content) but also a wide range of malleable iron with a carbon content below that of quenchable steel (0.5% carbon content). This malleable iron was often made directly into hoes, shovels, and other hand tools by a blacksmith, such as Joseph Jenks, whose forge was a component of the integrated ironworks at Saugus. See carbon content of iron and steel, malleable iron, muck bar, and wrought iron.

Warm-forging: The technique of forging high carbon edge tools and swords at temperatures below the critical temperature (723° C). Forging at higher temperatures causes the later embrittlement of steel at room temperature. See cold forging, hot- and warm-forging, and Wootz steel.

Water bellows: Japanese fan bellows run by a horizontal water wheel, later used in Korea and China; the main source of power for most Japanese forges in antiquity. See Wertime (1962).

Watered steel: Etched steel, displaying the pattern derived from the metallurgical composition of the steel and welding techniques of the smiths, including both their mechanical (hammering) and thermal (forging) treatment of the sword, knife, or gun barrel being constructed. The metal surface of the artifact being forged is then watered, i.e. treated by pickling, i.e. the submerging or covering its surface with such acidic substances as lemon juice, ferric sulfate, or "acidulated water" (Breant 1823). See, in particular, chapters 2-5, in Smith's (1960) classic, *A History of Metallography*. See Damascened steel, Japanese swords, and pattern-welding.

Water-hardened steels: Low carbon and low alloy steels that can only be hardened by quenching in water.

Water-powered drop hammer: Another term for a trip hammer. The drop hammer became the hydraulic helve hammer of the later 15th century. See trip hammer.

Water quenching: The one quenching method of sufficient speed to give full hardness to steels with a carbon content of 0.5 to 0.6%.

Weald of Sussex: The region southeast of London, which was, along with the Forest of Dean, an important iron and steel producing area of Tudor England during the age of the merchant adventurers. Manganese-laced siderite ores were mined in the Weald before 1600 (and then depleted) and may have attracted the attention of Roman smelters in the early Iron Age. See Cleere (1985), Forest of Dean, manganese, and ordnance.

Weardale: An important source of iron ore located in County Durham, England, in the late 17th and early 18th centuries for the brief florescence of the iron and steel industries in Newcastle and the Derwent Valley in northeast England, 1685 – 1750. Along with the Weald, a second location of manganese-bearing iron ores, Weardale ore was "…not greatly dissimilar from the Carinthian ores. From the pig iron smelted from such ores it would have been possible to produce 'natural steel' [German steel] by the continental method and it should be noted that there was a blast furnace operating on these ores at Allensford, nearby, from about 1670" (Barraclough 1984a, 64). Barraclough notes that shortly thereafter (1703) the Hollow Sword Blade Company was, along with most other forges in England, producing blister steel in a cementation furnace. See German steel, shear steel, and Weald of Sussex.

Weld: To soften by heat and join together two pieces of metal by hammering, pressure, or chemical reaction.

Welding: Using heat to join two metal parts together. The earliest ferrous welding technique is forge or hammer welding, in which the elements to be joined are shaped with heat and hammer on the anvil (scarfed) and then brought up to welding heat in the forge. The mating surfaces are treated with flux, often a preparation of borax and silica, to remove oxides from the joint, and then hammered, rolled, or rammed with sufficient force to allow the establishment of a molecular bond. Before the end of the 19th century, three new processes were emerging: oxy-acetylene or gas welding, resistance welding, and metal arc welding. Of these three, various forms of metal-arc welding have become the most widely used and the subject of the greatest refinement for industrial applications. Thus modern welding methods now include gas tungsten arc welding (GTAW) aka TIG, gas metal arc welding (GMAW) aka MIG, submerged arc welding (SAW), a mechanized MIG process, and shielded metal arc welding (SMAW) aka "stick" welding. It is this last technique that is most often referred to as arc welding, using the familiar welding rod, properly described as an electrode because it conducts electrical current. The process was advanced considerably by the invention in 1904 of the covered electrode. Before that, uncoated electrodes produced welds of uncertain properties. The flux coating serves in several capacities, stabilizing the arc, purifying the weld metal, and protecting the molten deposit from contact with oxygen and nitrogen in the air. More recent advances in welding strategies include electron beam welding, laser welding, and hybrid laser-arc welding, combining laser beam welding with GMAW, allowing deep penetration single pass butt welds in heavy gauge applications that ordinarily required multi-pass flux-cored arc welding (FCAW) (McLaughlin, David 2008, Liberty Salvage Co., personal communication).

Welding iron and steel: Joining two pieces of wrought iron, malleable iron, and/or low carbon steel at temperatures just above their melting point. See scarf weld.

Welding heat: The individual welding temperature necessary for a blacksmith to bring wrought iron and/or malleable iron and blister steel to their individual welding heats to facilitate "steeling," as in the forge welding of edge tools. The welding heat for wrought iron lies between 2200 – 2700° F (white heat). After reaching that temperature, wrought iron will give off sparks and droplets of melted metal. With increasing carbon content, the welding heat of steel drops, cast steel being the most difficult to weld. Steel containing carbon content in excess of 1.5% cannot be welded.

Weld steel: Term describing an edge tool produced by "steeling" or welding a steel bit or cutting edge on an iron shaft or haft. In the manufacturing of axes and other edge tools not made of cast crucible steel, a welded steel cutting edge (bit) was fused (welded) with the iron components of the tool, hence the name weld steel. Three strategies for welding steel cutting edges on iron shafted edge tools characterize ax and knife production since the early Iron Age. The most frequently used welding or "steeling" technique is the insertion of a steel bit between folded iron, especially for ax-making. The overcoat method of folding steel around an iron haft is a more recent 18th century ax-making technique. Scarf welding a steel cutting edge on the side of the iron body of larger offset broad axes is a third, less common technique of welding steel cutting edges on iron tools. Weld steel was produced in ancient times as natural steel and, after the advent of blast furnaces, as German steel (fined cast iron). The most common form of weld steel bar stock, used between 1700 and 1900, was blister steel derived from the carburization of wrought or malleable iron bar stock in the cementation furnace. Cast steel produced by the crucible steel process after 1750 may also have been used as weld steel. Most edge tools utilizing cast steel as weld steel are usually marked either "cast steel" or "warranted cast steel." Blacksmiths working before the development of the cementation furnace (1650) may have spotted nodules of natural steel accidentally produced in the iron-smelting process, extracted these fragments of steel from the bloom, and welded them onto iron sockets and ax bodies. See blister steel, bit, cast steel, cementation furnace, forge welding, scarf welding, steeling, and welding.

Whalecrafters: The New Bedford community of toolmakers who specialized in making harpoons, as well as edge tools, for the whaling communities of southern New England (Lytle 1984). The impost records of the New Bedford Customs District, made available by the New Bedford Whaling Museum, indicate that large quantities of Swedish iron were imported between 1816, the first year records were kept, and the 1830s to New Bedford. If not used to make the ductile wrought iron shafts of harpoons, Swedish bar iron was carburized in as yet undocumented steel (cementation) furnaces to make blister and shear steel for local edge toolmakers and for whalecrafters who also made steeled edge tools. After 150 years of making blister steel in England and a century of use in America, small steel furnaces were a likely component of the 19th century ironworks of many whalecrafters and edge toolmakers. See cementation furnace and edge tools.

Wheel cutting engine: A hand-operated machine for cutting teeth and gears; first used in England c. 1670.

Wheellock: A variation of the matchlock gun first produced in 1515 in Nuremburg. The wheellock was characterized by a flash hole with pyrites. The expensive-to-produce wheellock was particularly popular with the gentry in Elizabethan England, who used it for hunting in preference to the unwieldy matchlock arquebus. The wheellock was also the firearm of choice of continental royalty, nobility, and high-ranking military officers. The exquisite steel wheellock produced by Austrian-Nuremberg gunsmiths during the German Renaissance < 1635 was never surpassed in quality and beauty by any other firearm producing community. The advantage of the costly wheellock was that it could be fired with one hand. See Arquebus, flintlock, gun barrel iron, gunsmithing, and matchlock.

White cast iron: "Cast iron having cementite as its principal constituent" (Gordon 1996, 311). Cementite is iron chemically combined with graphite; it therefore contains little or no free graphite and has very limited commercial application unless it is melted and annealed into more useful grey cast iron. White cast iron is low in silicon and thus easier to convert to wrought iron than grey cast iron and, consequently, was the preferred iron used in the puddling furnace for conversion into wrought iron. It is produced by rapid cooling in an iron rather than sand mold and can then be annealed by heating to a cherry red heat for 60 hours to allow crystalline graphite to become free carbon, thereby producing malleable cast iron. The absence of graphite gives the fracture of white cast iron a white color. See cast iron, grey cast iron, malleable cast iron, and silicon.

White heart: An uncommon variant form of malleable cast iron in which all the free carbon within the interior of the casting is converted to chemically combined carbon during the annealing process producing a hard alloy (Spring 1917, 209-10).

White metal: Metal alloys with a significant tin content, which promotes elasticity and assists the even distribution of stress loads.

White metal alloy: Babbit alloys used in bearing composition in engines, motors, and turbines. Originally composed of tin ($\pm89\%$), copper ($\pm3.5\%$) and antimony ($\pm3\%$), modern bearings composites now include extensive use of lead, aluminum, and bronze-based alloys.

Whitesmith: Metalsmiths who used alloys of iron, zinc, and tin to galvanize metal products and make domestic utensils, such as teapots, usually by rolling, crimping, or otherwise shaping their wares; also called a tinsmith. See swage block.

Whitworth steel: A form of fluid, compressed steel that Sir Joseph Whitworth (c. 1862) adapted from Huntsman's technique of producing cast steel in crucibles. Its brand name was Wheatsheaf, and it was preferred for making shotgun barrels. Whitworth steel was later made by the Siemens-Martin (c. 1870) process. See cast steel.

Whitworth thread: The standard screw thread used in England and continental Europe, which has a 55° angle; named after its inventor, Sir Joseph Whitworth (1792-1871). See the discussion of Whitworth's role as an Industrial Revolutionary in *Hand Tools in History: Steel- and Toolmaking Strategies and Techniques before 1870* (Brack 2008).

Widmanstätten structure: The patterns formed by alternating areas of pearlite and ferrite after rapid cooling of low carbon steel containing large austenite grains. Alois von Widmanstätten first noted this crystal structure in 1808 on meteorites in the royal mineral collection in Vienna; the term now refers to artificial structures in annealed steels (Smith 1960, 150) characterized by ferritic/pearlitic plate morphology (Sinha 2003). See austenite, crystal structure, and Neumann bands.

Wilkinson, John: In the early 1750s, John Wilkinson invented a boring machine to manufacture precision ground cylinders, which were quickly adapted by Bolton and Watt in the manufacture of steam engines.

Wire annealing: The softening of cold-hardened or drawn wire by slow heating.

Wootz steel: Another name for Damascus steel, which originated in India in ancient times and was used for edge tools, swords, and sitar wire (Craddock 1985). An early form of crucible steel, it was made in +/- 3 kg batches in crucibles, especially in Muslim communities since at least the early Christian era. Bealer (1976a) suggests Wootz steel was made with molten cast iron in a manner similar to the Brescian method and was composed of a mixture of granules of soft iron and carbon steel. Moxon (1975) noted Wootz steel as one of the principal forms of steel available to toolmakers in the late 17th century. Smith (1960) indicates that Wootz steel was made in cakes, which were not fully melted, "but sometimes consisted of an aggregate of unmelted iron granules or little plates cemented together by once molten cast iron. Such a structure would result from the standard Indian process of steel making if the temperature were insufficient to melt the alloy completely. The crucible charge was an iron sponge, sometimes roughly forged into plates, together with wood for carbonization" (Smith 1960, 21). See crucible steel, Damascus steel, Moxon, superplasticity, Venetian steel, and Viking sword.

Wootz steel ingot: Ingots produced as disks usually weighing ±2 lb., 4" in diameter, 2" high, and widely traded to Persia and other western Asian locations (Buchanan 1807).

Wootz steel production process: The exact details of the Wootz steel production process are unknown, but two strategies have been suggested: 1) the fusion of cast and wrought iron (Lowe 1990) or 2) the carburization of wrought iron in crucibles packed with carboniferous materials. With either strategy, ±15 small crucibles were laid in an ash-filled pit adjacent to a fire pit and buffalo hide bellows provided the blast. Sanskrit texts written between the 7th and 13th centuries provide a description of the process. The smelting duration was approximately 24 hours at temperatures ranging from 1300° to 1400° C or higher, which produced a hypereutectoid steel (cc > 0.83%,) and, in some cases, an ultra high carbon steel (UHCS), with carbon contents ranging from 1% to as high as 2%. Wootz steel was used in the production of Damascus swords. The superplasticity of Wootz steel may be due to a combination of unique forging techniques and the presence of vanadium as a microconstituent (cobalt is also noted) within the nanotubes of the steel microstructure. The contemporary difficulty in reproducing Wootz steel may be due to the unavailability of ores containing these microconstituents (Srinivasan n.d.).

Work: "The amount of energy expended when a force is moved through a distance" (Maiklem 1998, 26).

Work hardening: Hardening of iron and steel by deformation that occurs during hammering, rolling, drawing, punching, and/or bending of metals at temperatures below the critical point. See slip interference and stress fields.

Wrought iron: Slag-bearing, malleable iron with low carbon content (0.08 % or less, but usually within the range of 0.02 - 0.04%) (Aston 1939). Wrought iron does not harden when cooled suddenly (quenching). Its two principal components are high purity iron (ferrite) and iron silicate, a glass-like slag with a high silicon content, which is present in a one inch wrought iron rod in the amount of 250,000 filaments (Aston 1939). Due to its malleability and ductility, it is easily shaped, threaded, machined, hammered, or stretched when hot. Its iron silicate content (siliceous slag) gives it unique characteristics, particularly its resistance to stress concentration (toughness and tensile strength), as well as its corrosion resistance. Produced in larger quantities with better quality control after the invention of the puddling (reverbatory) furnace in 1784, the purer wrought iron thus produced was the key ingredient in producing higher quality cementation and crucible steel. The slag inclusions in wrought iron aid its corrosion resistance, making it especially useful in stressful environments where moisture would quickly corrode steel. When forged across the grain instead of along the grain, loss of strength occurs. Knight (1875) provides this description of the refining of high quality wrought iron: "Iron sufficiently pure to be drawn out into bars and welded. . . The iron obtained from [simple furnaces] was a spongy mass, which was drawn out at intervals by tongs, hammered gently and cautiously, and then more energetically, to press the particles of iron into contact and squeeze out the intermixed slag. The mass thus obtained was cut up into smaller pieces, which were repeatedly heated and hammered till a sound bar of nearly pure iron was obtained" (Knight 1875, 1377). Utilizing slightly altered methods of refining also produced malleable iron with a higher carbon content, as well as wrought steel with the same carbon content as low carbon steel. Malleable iron and wrought steel still contained significant trace amounts of siliceous slag but, because of the slightly higher carbon content, were much more suitable for forging common tools, such as shovels, hoes, tongs, augers, leg vises, and early wagon wrenches, than softer and more ductile wrought iron, which was more easily deformed. See bloomery, direct process, iron silicate, malleable iron, muck bar, puddling furnace, and silicon.

Wrought iron and steel sword: The pattern-welded wrought iron and steel sword was a common form of high quality fighting sword from the early Iron Age to the late 19th century. The repeated forge welding of interspersed thin layers of tough, ductile, silicon-filament-laced wrought iron between thin layers of sheet steel provided unique qualities of durability and resistance to fracture. Japanese, Merovingian, and other swordmakers were experts at using this technique, which is still utilized in the modern age by samurai swordmakers. The all steel (?) swords of the Vikings and the Wootz steel swords of Damascus are variations of the usual method of pattern-welding swords. The all malleable iron sword, also a common form of weapon used during the early Iron Age until the late medieval period, was a distinctly inferior product. The quality of all three variations of swordmaking techniques was intimately connected to the skill of the swordsmith and the number of times each weapon was re-piled and reforged. See Damascus sword and Japanese samurai sword.

Wrought iron production: Direct process smelting of wrought iron occurs in stages. In a typical shaft furnace, heating is first used to remove water, and then the ascending gases begin decomposing iron carbonates at a temperature of 500° C. In lower areas of the furnace, reduction, which converts the higher oxides Fe_3O_4 and Fe_2O_3 to the lower oxide FeO, begins at a temperature of about 750° C. The production of pure wrought iron in this idealized furnace thus has the following composition: $CO + FeO = Fe + CO_2$. Successful smelting of wrought iron is predicated upon maintenance of an adequate CO level in the furnace material. Rising temperatures in the wrought iron-smelting process at or above 900° C result in the formation of iron carbides, the principal component of raw steel and cast irons. In blast furnaces, which run much hotter than direct process bloomeries, the higher temperatures achieved (1600° C) enhance the production of carbon monoxide, the principal cause of excess carbide formation in a direct process smelting furnace. Examinations of early Iron Age furnace remains show significant deposits of iron carbide consistent with the variable but heterogeneous carbon content of most currency bars recovered from early Iron Age smelting furnaces or the smithies that reforged these currency bars. Tylecote (1987), provides this riff on the formation of iron carbides during direct process smelting, thus illustrating the difficulties of smelting pure wrought iron:

> At 900 C the carbon would be in solution in the austenite phase of the iron. Under these conditions the slag would be solid and low in iron as it would be in equilibrium with the high carbon iron in contact with it. Drops of high carbon iron (in some cases molten) would then fall down to the hotter parts of the furnace where they would be oxidized by the blast to lower the carbon content of the metal and raise the iron content of the slag in equilibrium with it:

$$Fe_3C + O_2 \rightarrow 3Fe + CO_2$$

$$2Fe + O_2 + SiO_2 \rightarrow 2FeO.SiO_2$$

Equilibrium is not usually achieved, certainly not over the whole cross-section of the furnace, so that the result is a rather heterogeneous mixture of high- and low-carbon areas with an average carbon level that is low. With such a carbon level the iron is solid (at 1200 C) but the slag becomes molten and runs away leaving a solid iron bloom with some porosity.

This process can continue until the space between the tuyere and the bottom is full of metal, slag and charcoal. When this occurs the furnace must be opened or the tuyere raised and the bloom removed from the bottom. (Tylecote 1987, 151-152)

Tylecote is discussing bloomsmithing in an early Iron Age shaft furnace. Production of pure wrought iron in such a facility was extremely difficult. Most surviving currency bars have a heterogeneous carbon content, which would include areas of wrought iron with a low carbon content, as well as malleable iron and raw steel with a higher carbon and iron carbide content.

Wrought steel: A high carbon form of wrought iron produced in the puddling furnace or in fineries, later replaced by Bessemer and Siemens-Martin's open-hearth low carbon steel. Wrought steel is, in essence, a malleable iron with a higher carbon content than wrought iron and was forged out of solid iron bar stock. Many leg vises are composed of wrought steel and were probably made by a blacksmith who specialized in toolmaking. Wrought steel is also a modern generic term for any rolled or forged steel product, obscuring its older meaning.

Wüstite: A naturally occurring mineral that is also produced by the smelting of iron in a highly reducing environment, which, when liquid, is characterized by slowly growing crystal microstructures.

Yaki-ire: The rapid cooling by water quenching of Japanese steel swords after heating to the critical temperature (T_c: 1341° F / 723° C), characterized by the appearance of a bright red-orange color best observed in the darkened environment of the edge toolmaker or swordsmith. In the case of Japanese sword production, the thickness of the clay enclosure covering the steel sword during the quenching process determined the cooling rate and thus the microstructure of the steel sword. See quench, naganita, tsuchioki, and yaki-modoshi.

Yaki-modoshi: The further heat treatment of the hard sword blades (tempering) to a temperature at least 300° F less than the critical temperature, followed by re-quenching to relieve internal stresses. The degree of hardness or softness of Japanese swords after rapid quenching was dependent on the thickness of the clay enclosure during the tempering process. Sword edges covered by a thin layer of clay cooled more quickly and were thus harder than the softer and more flexible sword bodies. Similar techniques would have been used by shipsmiths and edge toolmakers who knew the art of edge tool forging. See kan of the ironmonger, temper, tsuchioki, and yaki-ire.

Yield point: The load per unit area at which deformation or elongation occurs without further increase in stress.

Zelechovice furnace: Slavic bloomery furnace that was used in the late 8th and early 9th centuries in the Moravian section of Eastern Europe. Of particular interest with respect to the issue of direct process natural steel production, these primitive, bowl-type furnaces were built into the ground and were characterized by a unique bowl cavity at the bottom of the furnace, in which the bloom of iron was protected from the oxidizing effect of burning charcoal, while at the same time being covered by the charcoal. Leaving the bloom of iron within the bowl cavity at the bottom of the furnace adjacent to both the tuyère and the burning charcoal produced heterogeneous blooms of natural steel. See Pleiner (1969, 461) for an illustration of the basic design of this early furnace, also reproduced in volume 6 of the *Hand Tools in History* series: *Steel- and Toolmaking Strategies and Techniques before 1870* (Brack 2008).

Appendix I: A Guide to the Metallurgy of the Edge Tools at the Davistown Museum

I. Steelmaking Strategies 1900 BC – 1930 AD

1. Natural Steel: 1900 BC – 1930 AD

Natural steel was made in direct process bloomeries, either deliberately or accidentally, in the form of occasional nodules of steel (+/- 0.5% carbon content (cc)) entrained in wrought iron loups. Bloomsmiths deliberately made natural steel for sword cutlers by altering the fuel to ore ratio in the smelting process, producing heterogeneous blooms of malleable iron (0.08 to 0.2% cc) and/or natural steel (0.2 to 0.5 cc and higher) or by carburizing bar or sheet iron submerged in a charcoal fire. Manganese-laced rock ores (e.g. from Styria in Austria or from the Weald in Sussex, England) facilitated natural steel production. As a slag constituent, manganese lowered the melting temperature of slag, facilitating the more uniform uptake of carbon in the smelted iron. The Chalybeans produced the first documented natural steel at the height of the Bronze Age in 1900 BC, using the self-fluxing iron sands from the south shores of the Black Sea. Occasional production of bloomery-derived natural steel edge tools continued in isolated rural areas of Europe and North America into the early 20th century.

2. German Steel: 1350 - 1900

German steel was produced by decarburizing blast-furnace-derived cast iron in a finery furnace, and, after 1835, in a puddling furnace. German steel tools are often molded, forged, or cast entirely of steel, as exemplified by trade and felling axes without an inserted (welded) steel bit. Such tools were a precursor of modern cast steel axes and rolled cast steel timber framing tools. German steel shared the world market for steel with English blister and crucible steel until the mid-19th century.

3. Blister Steel: 1650 - 1900

Blister steel was produced by carburizing wrought iron bar stock in a sandstone cementation furnace that protected the ore from contact with burning fuel. It was often refined by piling, hammering, and reforging it into higher quality shear or double shear steel or broken up and remelted in crucibles to make cast steel. Blister steel was often used for "steeling" (welding on a steel cutting edge or bit) on axes and other edge tools.

4. Shear Steel: 1700 - 1900

Shear steel was made from refined, reforged blister steel and used for "steeling" high quality edge tools, such as broad axes, adzes, and chisels, especially by American edge toolmakers who did not have access to, or did not want to purchase, expensive imported English cast steel. The use of shear steel was an alternative to imported English cast steel for making edge tools in America from the late 18th century to the mid-19th century.

5. Crucible Cast Steel: 1750 - 1930

Crucible cast steel was made from broken up pieces of blister steel bar stock, which was inserted into clay crucibles with small quantities of carboniferous materials (e.g. charcoal powder). After melting at high temperatures, crucible cast steel was produced in 5 to 25 kg batches and considered to be the best steel available for edge tool, knife, razor, and watch spring production. Due to lack of heat resistant clay crucibles, extensive production of high quality crucible cast steel didn't begin in the United States until after the Civil War.

6. Brescian Steel: 1350 - 1900?

Brescian Steel was a common Renaissance era strategy used in southern Europe to make, for example, steel for the condottiers of the Italian city states. Wrought or malleable iron bar stock was submerged and, thus, carburized in a bath of molten pig iron. Brescian steel cannot be visually differentiated from German steel or puddled steel, both of which were produced from decarburizing pig iron.

7. Bulk Processed Steel: 1870 f.

After the American Civil War, a number of new strategies were invented for producing large quantities of steel, especially low carbon steel, which was required by the rapid growth of the industrial age and its factory system of mass production. The first important innovation was Henry Bessemer's single step hot air blast process, followed by several variations of the Siemens-Martin open-hearth furnace and electric arc furnaces. For edge tool production, the electric arc furnace supplanted, and then replaced, crucible cast steel in the early decades of the 20th century. A few modern drop-forged edge tools are included in this exhibition as examples of modern bulk process steel producing strategies.

Appendix II: Art of the Edge Tool

I. Edge Toolmaking Techniques 1900 BC – 1930 AD

Shaping and Forging by Hand

A. Forge welding: Edge carburizing by heating, followed by hammering and additional heat treatment

B. Steeling: The welding of a steel bit onto an iron shaft or body

C. Pattern-welding: The welding together of alternating layers of sheet iron and steel, used by knife and swordmakers; seldom used by edge toolmakers

D. Molding: The shaping of short lengths of hot malleable iron or German steel bar stock in an iron pattern; sometimes the iron pattern was water-cooled. This method was not used after blister steel became widely available around 1700.

Shaping and Forging by Machine

E. Rolling: The hot rolling of cast steel into bar stock, and its further shaping by the formation of sockets, grinding, and further forging, both before and after additional thermal treatment.

F. Casting: The hot rolling of cast steel into steel bar stock compatible with its further shaping in molds or patterns by drop-forging, as in the drop-forging of cast steel axes.

G. Drop-forging: The hydraulic pressing of low carbon steel and malleable iron into tool forms by using dies as patterns as in the mass production of factory-made hand tools. Also, its casting in special purpose molds for the production of machinery and equipment of every conceivable use.

Most hand tools made in the 20th century show no evidence of hand work, but, in a minority of cases, (e.g. the ax) there is no clear distinction between the hand-forged and the machine-made tool until the late 20th century. Most edge tools made before 1930 are "hand-forged" or "forge welded" to some extent, no matter the technique used to "steel" their edges. The trip hammer and the water wheel are examples of machines that assisted edge toolmakers in the forging of their tools. The advent of the modern rolling mill (Henry Cort, 1784) for hot rolling cast steel bar stock did not end the long tradition of hand-forging an edge tool. When the Collins Ax Factory began drop-forging all steel axes sometime after 1837, many smaller ax companies continued hand-forging and hand hammering axes they produced, often with the aid of other machinery, well into the 20th century. The evolution from hand-forging to machine forging (drop-forging) hand tools was thus a gradual process. One goal of the creative economy of the post-industrial era is the revival of handmade hand toolmaking strategies and techniques.

Appendix III: Classification for Metals and Alloys

Table 5. Unified Numbering System (UNS) for Metals and Alloys. From Oberg et al. ([1914] 1996, 402) *Machinery's Handbook: 25th edition: A reference book for the mechanical engineer, designer, manufacturing engineer, draftsman, toolmaker, and machinist.*

UNS Series	Metal
A00001 to A99999	Aluminum and aluminum alloys
C00001 to C99999	Copper and copper alloys
D00001 to D99999	Specified mechanical property steels
E00001 to E99999	Rare earth and rare earthlike metals and alloys
F00001 to F99999	Cast irons
G00001 to G99999	AISI and SAE carbon and alloy steels (except tool steels)
H00001 to H99999	AISI and SAE H-steels
J00001 to J99999	Cast steels (except tool steels)
K00001 to K99999	Miscellaneous steels and ferrous alloys
L00001 to L99999	Low-melting metals and alloys
M00001 to M99999	Miscellaneous nonferrous metals and alloys
N00001 to N99999	Nickel and nickel alloys
P00001 to P99999	Precious metals and alloys
R00001 to R99999	Reactive and refractory metals and alloys
S00001 to S99999	Heat and corrosion resistant (stainless) steels
T00001 to T99999	Tool steels, wrought and cast
W00001 to W99999	Welding filler metals
Z00001 to Z99999	Zinc and zinc alloys

Appendix IV: Compositions of AISI-SAE Standard Carbon Steels

Table 6. From Oberg et al. ([1914] 1996, 407-8) Machinery's Handbook: 25th edition: A reference book for the mechanical engineer, designer, manufacturing engineer, draftsman, toolmaker, and machinist.

AISI NO.	Composition (%)				SAE No.
	C	Mn	P Max.	S Max.	
Nonresulfurized Grades – 1 per cent Mn (max)					
1005	0.06 max	.035 max	.040	.050	1005
1006	0.08 max	0.25-0.40	.040	.050	1006
1008	0.10 max	1.30-0-50	.040	.050	1008
1010	0.08-0.13	0.30-0.60	.040	.050	1010
1011	0.08-0.13	0.60-0.90	.040	.050	—
1012	0.10-0.15	0.30-0.60	.040	.050	1012
1013	0.11-0.16	0.50-0.80	.040	.050	1013
1015	0.13-0.18	0.30-0.60	.040	.050	1015
1016	0.13-0.18	0.60-0.90	.040	.050	1016
1017	0.15-0.20	0.30-0.60	.040	.050	1017
1018	0.15-0.20	0.60-0.90	.040	.050	1018
1019	0.15-0.20	0.70-1.00	.040	.050	1019
1020	0.10-0.23	0.30-0.60	.040	.050	1020
M1020	0.17-0.24	0.25-0.60	.040	.050	—
1021	0.18-0.25	0.60-0.90	.040	.050	1021
1022	0.18-0.23	0.70-1.00	.040	.050	1022
1023	0.20-0.25	0.30-0.60	.040	.050	1023
1025	0.22-0.28	0.30-0.60	.040	.050	1025
1026	0.22-0.28	0.60-0.90	.040	.050	1026
1029	0.25-0.31	0.60-0.90	.040	.050	—
1030	0.28-0.34	0.60-0.90	.040	.050	1030
1034	0.32-0.28	0.50-0.80	.040	.050	—
1035	0.32-0.38	0.60-0.90	.040	.050	1035
1037	0.32-0.38	0.70-1.00	.040	.050	1037
1038	0.35-0.42	0.60-0.90	.040	.050	1038

AISI NO.	Composition (%)				SAE No.
	C	Mn	P Max.	S Max.	
1039	0.37-0.44	0.70-1.00	.040	.050	1039
1040	0.37-0.44	0.60-0.90	.040	.050	1040
1042	0.40-0.47	0.60-0.90	.040	.050	1042
1043	0.40-0.47	0.70-1.00	.040	.050	1043
1044	0.43-0.50	0.30-0.60	.040	.050	1044
M1044	0.40-0.50	0.25-0.60	.040	.050	—
1045	0.43-0.50	0.60-0.90	.040	.050	1045
1046	0.43-0.50	0.70-1.00	.040	.050	1046
1049	0.46-0.53	0.60-0.90	.040	.050	1049
1050	0.48-0.55	0.60-0.90	.040	.050	1050
M1053	0.48-0.55	0.70-1.00	.040	.050	—
1055	0.50-0.60	0.60-0.90	.040	.050	1055
1059	0.55-0.65	0.50-0.80	.040	.050	1059
1060	0.55-0.65	0.60-0.90	.040	.050	1060
1064	0.60-0.70	0.50-0.80	.040	.050	1064
1065	0.60-0.70	0.60-0.90	.040	.050	1065
1069	0.65-0.75	0.40-0.70	.040	.050	1069
1070	0.65-0.75	0.60-0.90	.040	.050	1070
1071	0.65-0.70	0.75-1.05	.040	.050	—
1074	0.70-0.80	0.50-0.80	.040	.050	1074
1075	0.70-0.80	0.40-0.70	.040	.050	1075
1078	0.72-0.85	0.30-0.60	.040	.050	1078
1080	0.75-0.88	0.60-0.90	.040	.050	1080
1084	0.80-0.93	0.60-0.90	.040	.050	1084
1086	0.80-0.93	0.30-0.50	.040	.050	1086
1090	0.85-0.98	0.60-0.90	.040	.050	1090
1095	0.90-1.03	0.30-0.50	.040	.050	1095
Nonresulferized Grades – Over 1 per cent Mn					
1513	0.10-0.16	1.10-1.40	.040	.050	1513
1518	0.15-0.21	1.10-1.40	.040	.050	—
1522	0.18-0.24	1.10-1.40	.040	.050	1522

AISI NO.	Composition (%)				SAE No.
	C	Mn	P Max.	S Max.	
1524	0.19-0.25	1.35-1.65	.040	.050	1524
1525	0.23-0.29	0.80-1.10	.040	.050	—
1526	0.22-0.29	1.10-1.40	.040	.050	1526
1527	0.22-0.29	1.20-1.50	.040	.050	1527
1536	0.30-0.37	1.20-1.50	.040	.050	1536
1541	0.36-0.44	1.35-1.65	.040	.050	1541
1547	0.43-0.51	1.35-1.65	.040	.050	—
1548	0.44-0.52	1.10-1.40	.040	.050	1548
1551	0.45-0.56	0.85-1.15	.040	.050	1551
1552	0.47-0.55	1.20-1.50	.040	.050	1552
1561	0.55-0.65	0.75-1.05	.040	.050	1561
1566	0.60-0.71	0.85-1.15	.040	.050	1566
1572	0.65-0.76	1.00-1.30	.040	.050	—
Free-Machining Grades - Resulfurized					
1108	0.08-0.13	0.50-0.80	.040	0.08-0.13	1108
1109	0.08-0.13	0.60-0.90	.040	0.08-0.13	1109
1110	0.08-0.13	0.30-0.60	.040	0.08-0.13	1110
1116	0.14-0.20	1.10-1.40	.040	0.16-0.23	1116
1117	0.14-0.20	1.00-1.30	.040	0.08-0.13	1117
1118	0.14-0.20	1.30-1.60	.040	0.08-0.13	1118
1119	0.14-0.20	1.0-1.30	.040	0.24-0.33	1119
1132	0.27-0.34	1.35-1.65	.040	0.08-0.13	1132
1137	0.32-0.39	1.35-1.65	.040	0.08-0.13	1137
1139	0.35-0.43	1.35-1.65	.040	0.13-0.20	1139
1140	0.37-0.44	0.70-1.00	.040	0.08-0.13	1140
1141	0.37-0.45	1.35-1.65	.040	0.08-0.13	1141
1144	0.40-0.48	1.35-1.65	.040	0.24-0.33	1144
1145	0.42-0.49	0.70-1.00	.040	0.04-0.07	1145
1146	042-0.49	0.70-1.00	.040	0.08-0.13	1146
1151	0.48-0.55	0.70-1.00	.040	0.08-0.13	1151

AISI NO.	Composition (%)				SAE No.
	C	Mn	P Max.	S Max.	
Free-Machining Grades – Resulfurized and Rephosphorized					
1211	0.13 max	0.60-0.90	0.07-0.12	0.10-0.15	1211
1212	0.13 max	0.70-1.00	0.07-0.12	0.16-0.23	1212
1213	0.13 max	0.70-1.00	0.07-0.12	0.24-0.33	1213
1215	0.09 max	0.75-1.05	0.04-0.09	0.26-0.35	1215
12L 14	0.15 max	0.85-1.15	0.04-0.09	0.26-0.35	12L14

Appendix V: Classification for Carbon and Alloy Steels

Table 7. AISI-SAE System of Designating Carbon and Alloy Steels. From Oberg et al. ([1914] 1996, 406) *Machinery's Handbook: 25th edition: A reference book for the mechanical engineer, designer, manufacturing engineer, draftsman, toolmaker, and machinist.*

SAE designation	Type
Carbon and alloy steel grades	
Carbon steels	
10xx	Plain carbon (Mn 1.00% max)
11xx	Resulfurized
12xx	Resulfurized and rephosphorized
15xx	Plain carbon (Mn 1.00% to 1.65%)
Manganese steels	
13xx	Mn 1.75%
Nickel steels	
23xx	Ni 3.50%
25xx	Ni 5.00%
Nickel-chromium steels	
31xx	Ni 1.25%, Cr 0.65% or 0.80%
32xx	Ni 1.25%, Cr 1.07%
33xx	Ni 3.50%, Cr 1.50% or 1.57%
34xx	Ni 3.00%, Cr 0.77%
Molybdenum steels	
40xx	Mo 0.20% or 0.25% or 0.25% Mo & 0.042 S
44xx	Mo 0.40% or 0.52%
Chromium-molybdenum (Chromoly) steels	
41xx	Cr 0.50% or 0.80% or 0.95%, Mo 0.12% or 0.20% or 0.25% or 0.30%
Nickel-chromium-molybdenum steels	
43xx	Ni 1.82%, Cr 0.50% to 0.80%, Mo 0.25%
43BVxx	Ni 1.82%, Cr 0.50%, Mo 0.12% or 0.35%, V 0.03% min
47xx	Ni 1.05%, Cr 0.45%, Mo 0.20% or 0.35%

Carbon and alloy steel grades	
SAE designation	**Type**
81xx	Ni 0.30%, Cr 0.40%, Mo 0.12%
81Bxx	Ni 0.30%, Cr 0.45%, Mo 0.12%
86xx	Ni 0.55%, Cr 0.50%, Mo 0.20%
87xx	Ni 0.55%, Cr 0.50%, Mo 0.25%
88xx	Ni 0.55%, Cr 0.50%, Mo 0.35%
93xx	Ni 3.25%, Cr 1.20%, Mo 0.12%
94xx	Ni 0.45%, Cr 0.40%, Mo 0.12%
97xx	Ni 0.55%, Cr 0.20%, Mo 0.20%
98xx	Ni 1.00%, Cr 0.80%, Mo 0.25%
Nickel-molybdenum steels	
46xx	Ni 0.85% or 1.82%, Mo 0.20% or 0.25%
48xx	Ni 3.50%, Mo 0.25%
Chromium steels	
50xx	Cr 0.27% or 0.40% or 0.50% or 0.65%
50xxx	Cr 0.50%, C 1.00% min
50Bxx	Cr 0.28% or 0.50%
51xx	Cr 0.80% or 0.87% or 0.92% or 1.00% or 1.05%
51xxx	Cr 1.02%, C 1.00% min
51Bxx	Cr 0.80%
52xxx	Cr 1.45%, C 1.00% min
Chromium-vanadium steels	
61xx	Cr 0.60% or 0.80% or 0.95%, V 0.10% or 0.15% min
Tungsten-chromium steels	
72xx	W 1.75%, Cr 0.75%
Silicon-manganese steels	
92xx	Si 1.40% or 2.00%, Mn 0.65% or 0.82% or 0.85%, Cr 0.00% or 0.65%
High-strength low-alloy steels	
9xx	Various SAE grades
xxBxx	Boron steels
xxLxx	Leaded steels

Appendix VI: Classification of Tool Steels

Table 8. From Oberg et al. ([1914] 1996, 454) *Machinery's Handbook: 25th edition: A reference book for the mechanical engineer, designer, manufacturing engineer, draftsman, toolmaker, and machinist.*

Category Designation	Letter Symbol	Group Designation
High-Speed Tool Steels	M	Molybdenum types
	T	Tungsten types
Hot-Work Tool Steels	H_1-H_{19}	Chromium types
	H_{20}-H_{39}	Tungsten types
	H_{40}-H_{59}	Molybdenum types
Cold-Work Tool Steels	D	High-carbon, high-chromium types
	A	Medium-alloy, air-hardening types
	O	Oil-hardening types
Shock-Resisting Tool Steels	S	…
Mold Steels	P	…
Special-Purpose Tool Steels	L	Low-alloy types
	F	Carbon-tungsten types
Water-Hardening Tool Steels	W	…

Appendix VII: Iron-Carbon Diagrams

The following iron-carbon diagrams reflect only a small portion of the wide variety of possible interpretations of iron-carbon alloy types and their relationship to carbon content, temperature, and cooling rate. See, in particular, the transformation time diagram.

Schematic iron-carbon diagram

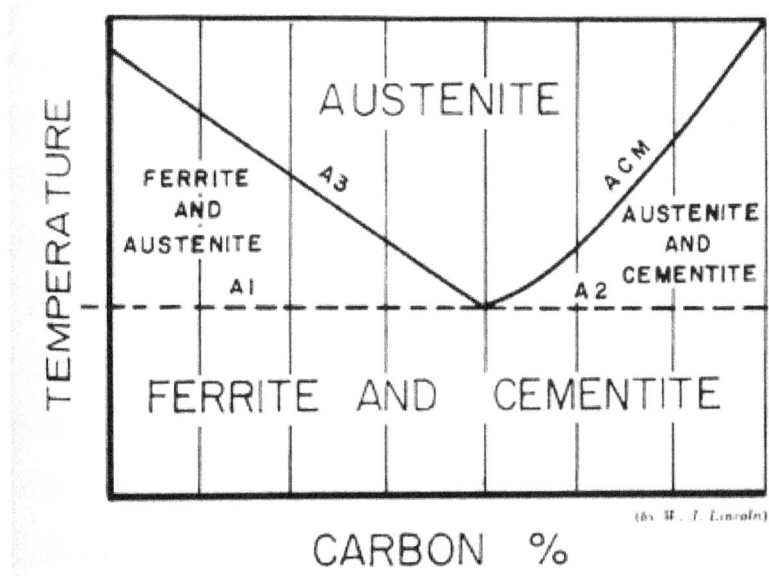

Figure 73. Schematic iron-carbon diagram showing areas important in heat treatment.

Figure 2 Shrager, M. 1961. *Elementary Metallurgy and Metalography*. NY: Dover. pg. 135. Used with permission of Dover.

Iron-carbon phase diagram

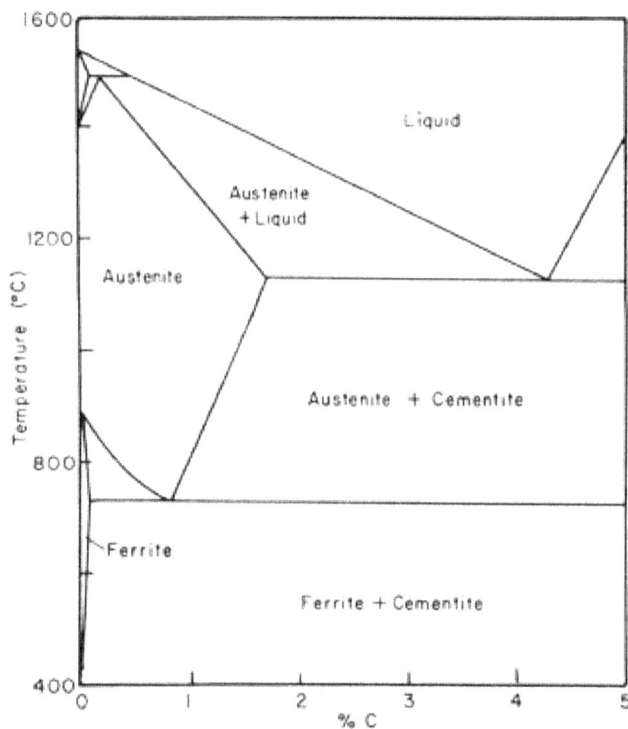

A-1. *The iron-carbon phase diagram shows the temperatures and compositions at which the different constituents of iron-carbon alloys are stable. Ordinarily, these constituents are austenite, ferrite, and cementite (iron carbide, Fe_3C). The constituent that Henry Sorby named pearlite consists of plates of cementite in a matrix of ferrite; it is formed by the decomposition of austenite at 723°C. If an alloy containing more than about 2.5 percent carbon is cooled very slowly, or if there is silicon in the iron, the carbon may appear as graphite rather than cementite.*

Figure 3 Gordon, Robert B. © 1996. *American Iron, 1607-1900.* pg. 252. The Johns Hopkins University Press. Reprinted with permission of The Johns Hopkins University Press.

Iron-carbon constitutional diagram 1

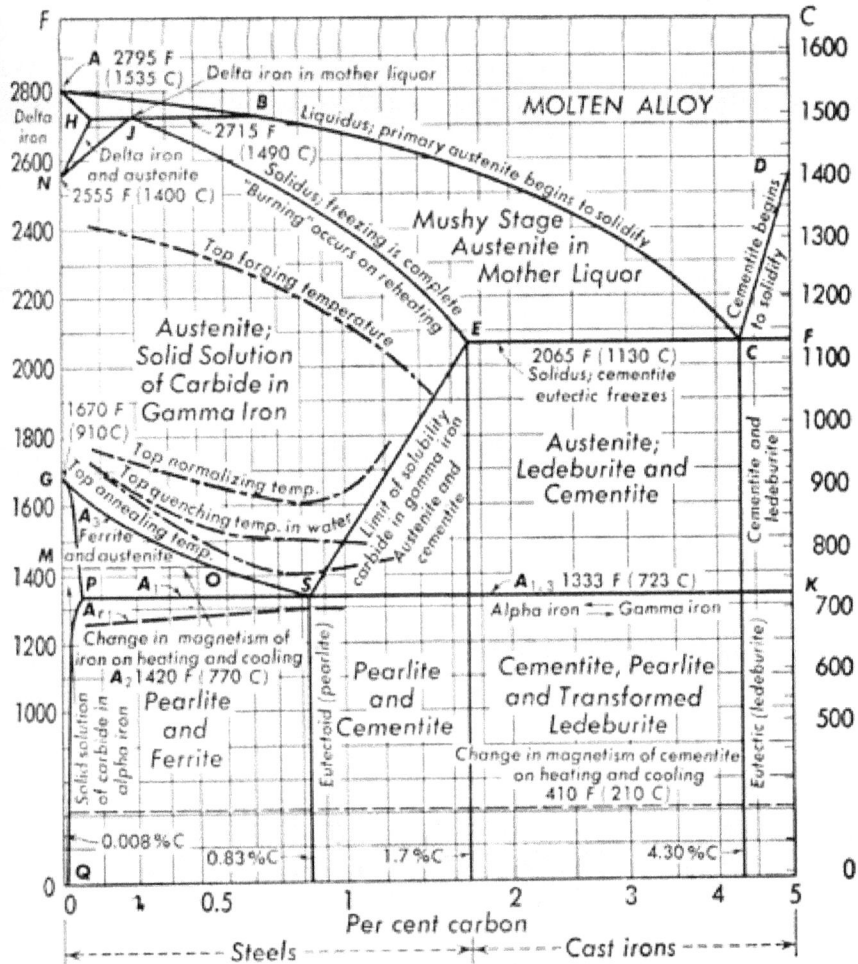

Figure 4 Shrager, M. 1961. *Elementary Metallurgy and Metalography*. NY: Dover. Figure 20. pg. 35. Used with permission of Dover.

Cooling curve of pure iron diagram

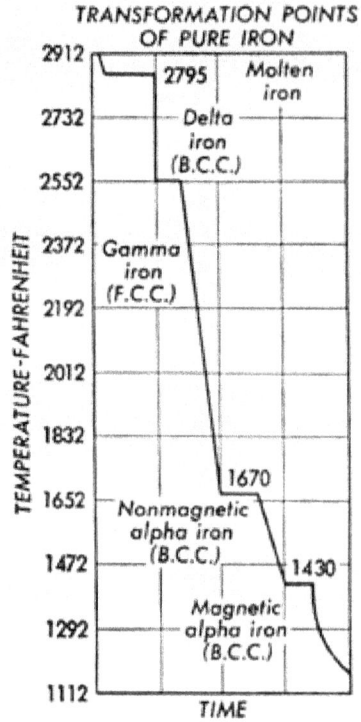

TRANSFORMATION POINTS
OF PURE IRON

Figure 19. Cooling curve of pure iron.

Figure 5 Shrager, M. 1961.
*Elementary Metallurgy and
Metalography.* NY: Dover. pg. 34.
Used with permission of Dover.

Solidus-liquidus diagram

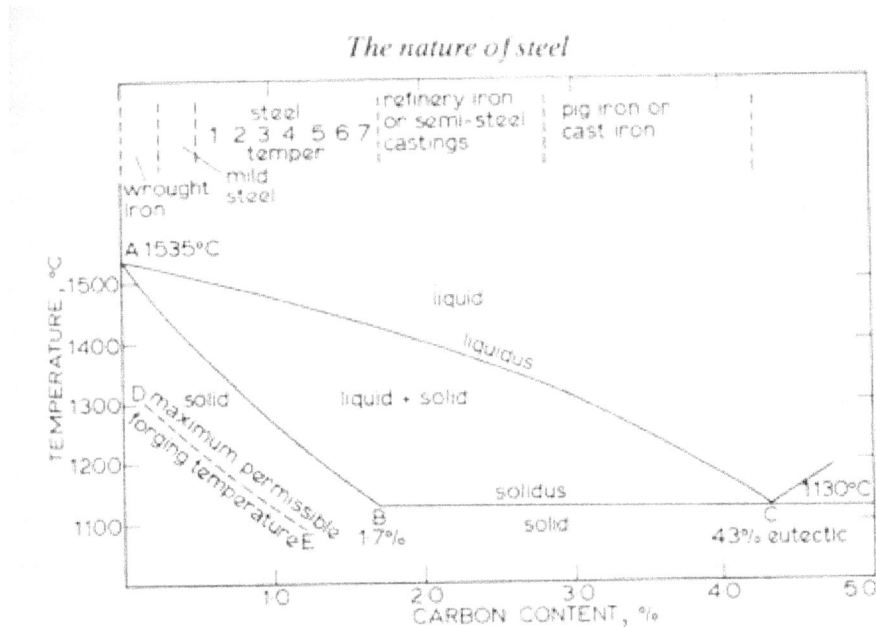

The nature of steel

Fig. 3   Effect of carbon content on melting and freezing point of iron

Pure iron freezes and melts at 1 535°C. An alloy of iron with 4·3% carbon freezes and melts at a temperature of 1 130°C; this is a 'eutectic' alloy. All other iron–carbon alloys melt and freeze over a temperature range. This is determined by the line *AC*, which is known as the *liquidus*, and the line *AB*, continued along *BC*, known as the *solidus*. On heating, when the solidus line is reached, partial melting begins, continuing as the temperature rises, and culminating in the alloy becoming completely liquid when the liquidus line is reached. The reverse occurs on cooling. The maximum permissible forging temperature lies below the solidus line, typically as indicated by the line *DE*. Alloys with more than 1·5% carbon are virtually unforgeable.

Also indicated are the classes of material as defined by carbon content, steel for the purposes of the current discussion being an alloy of iron containing 0·5–1·5% carbon.

Figure 6 Barraclough, K.C. 1984a. *Steelmaking before Bessemer: Blister steel, the birth of an industry*. Volume 1. The Metals Society, London, England. pg. 5. Used with permission from the Institute of Materials, Minerals and Mining.

Iron-ironcarbide diagram

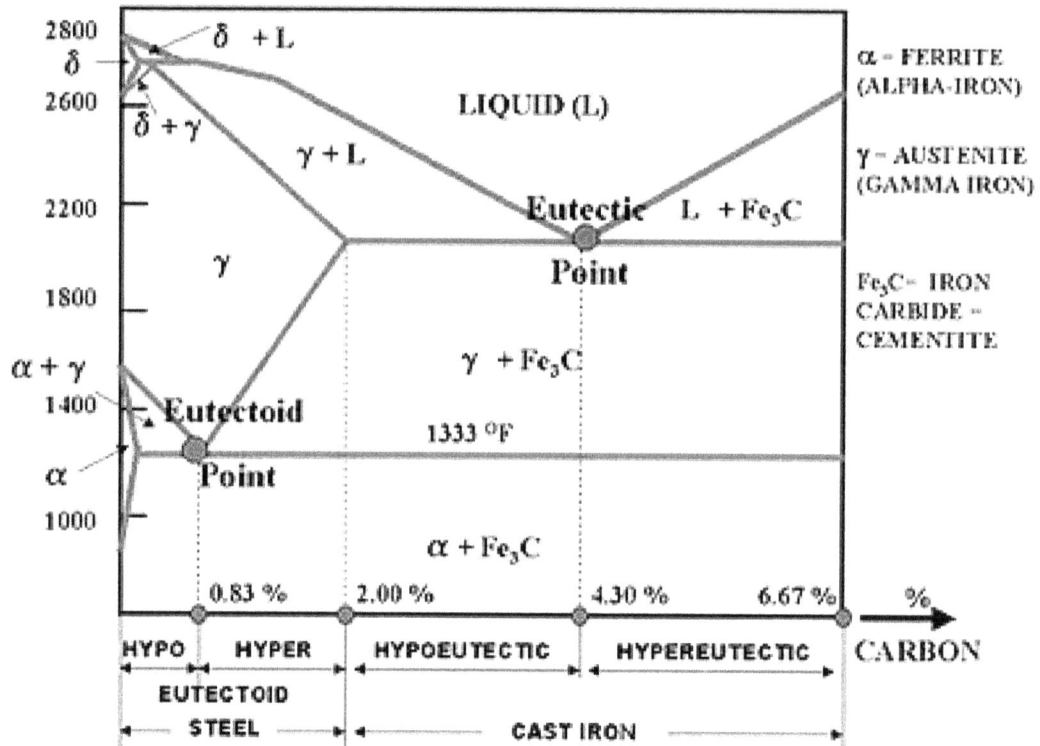

Figure 7 Used with permission from Serdar Z. Elgun.
http://info.lu.farmingdale.edu/depts/met/met205/fe3cdiagram.html

Critical temperature: heating versus cooling diagram

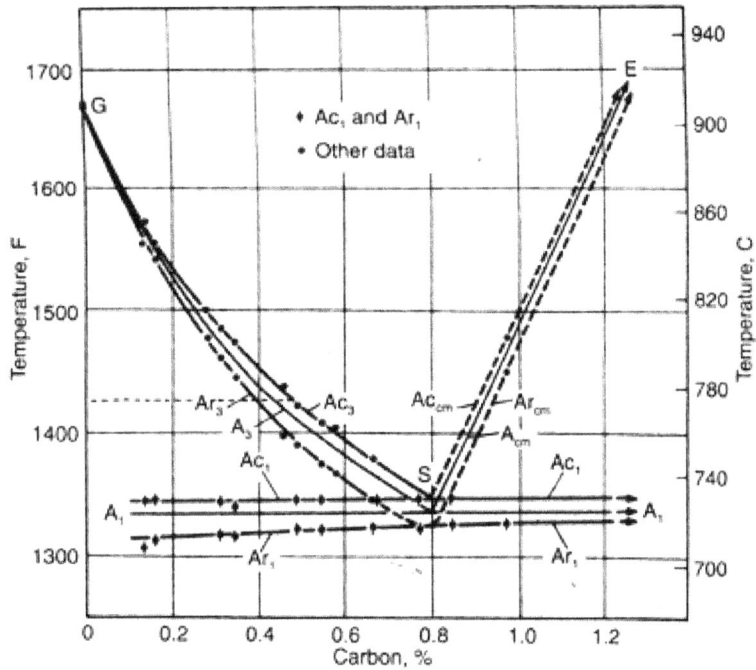

FIGURE 1.4 A portion of Fe-Fe₃C diagram showing two sets of critical cooling temperatures: Ac_1, Ac_3, and Ac_{cm} for heating and Ar_1, Ar_3, and Ar_{cm} for cooling. Rate of heating and cooling at 0.125°C/min.⁹ (*Reprinted by permission of ASM International, Materials Park, Ohio.*)

Figure 8 Sinha, Anil Kumar. 2003. *Physical metallurgy handbook*. New York: McGraw-Hill. pg. 1.8. Reproduced with permission of The McGraw-Hill Companies.

Alpha-gamma-cementite eutectoid diagram

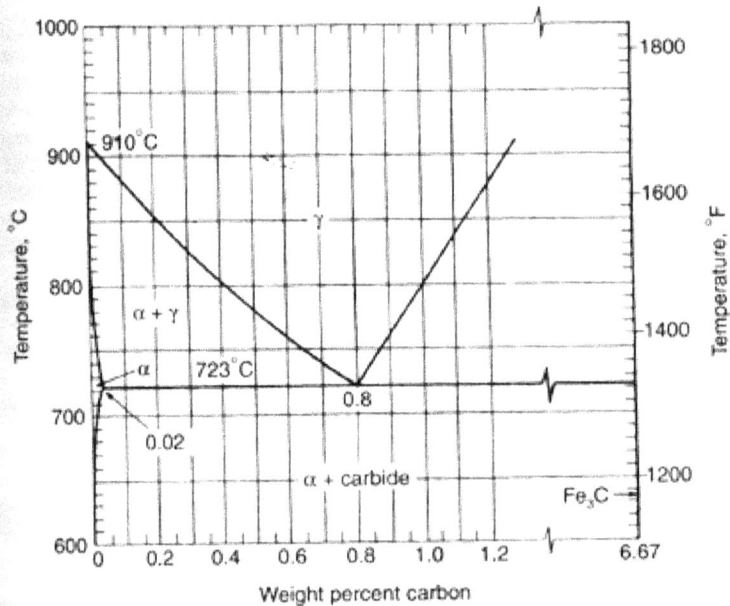

FIGURE 1.5 The eutectoid portion of the Fe-Fe₃C diagram. (*Reprinted by permission of Addison-Wesley Publishing Co., Reading, Massachusetts.*)

Figure 9 Sinha, Anil Kumar. 2003. Physical metallurgy handbook. New York: McGraw-Hill. pg. 1.9. Reproduced with permission of The McGraw-Hill Companies.

Iron-carbon constitutional diagram 2

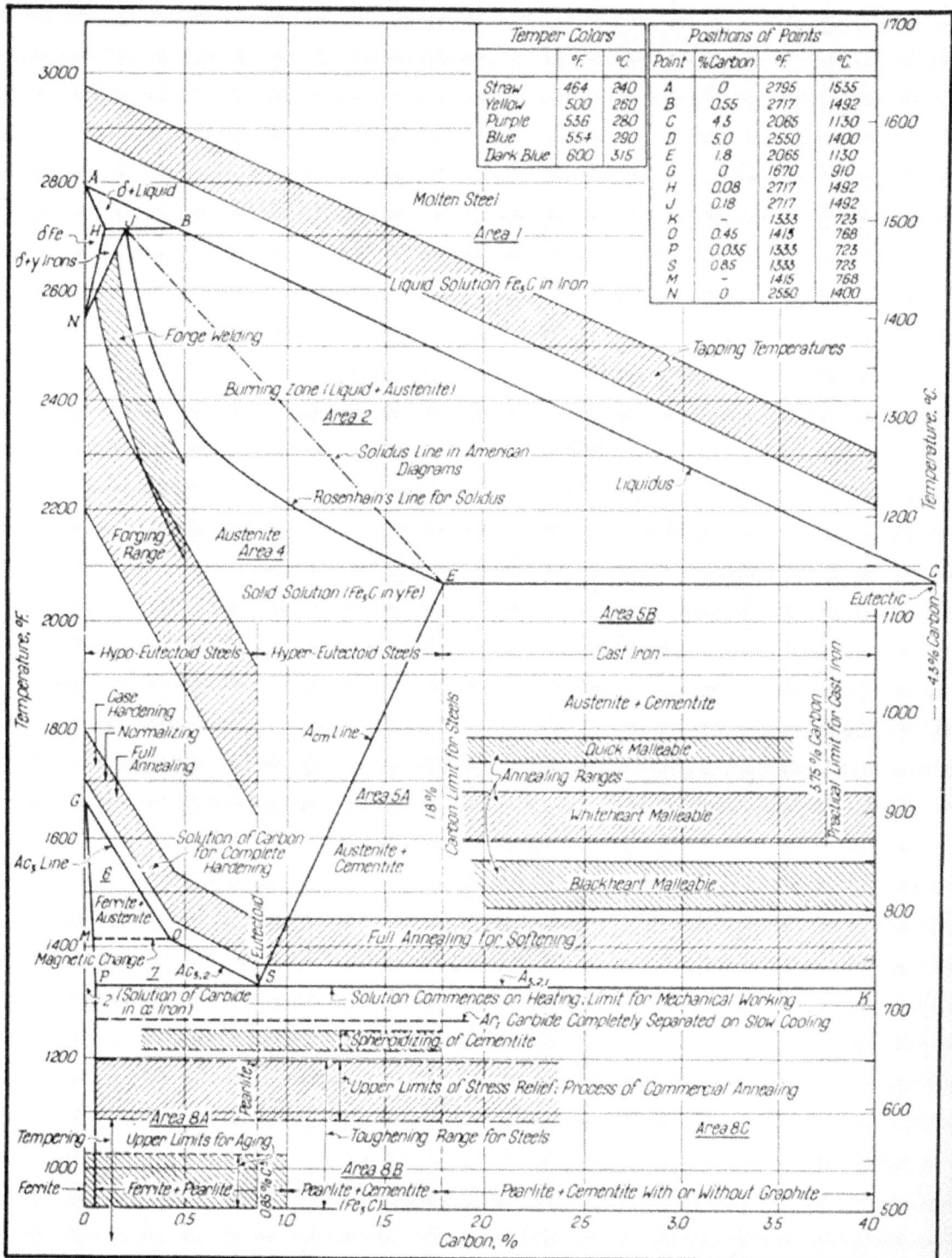

Figure 10 Shrager, M. 1961. Elementary Metallurgy and Metalography. NY: Dover. pg. 38. Reproduced with permission from Dover.

Transformation time diagram

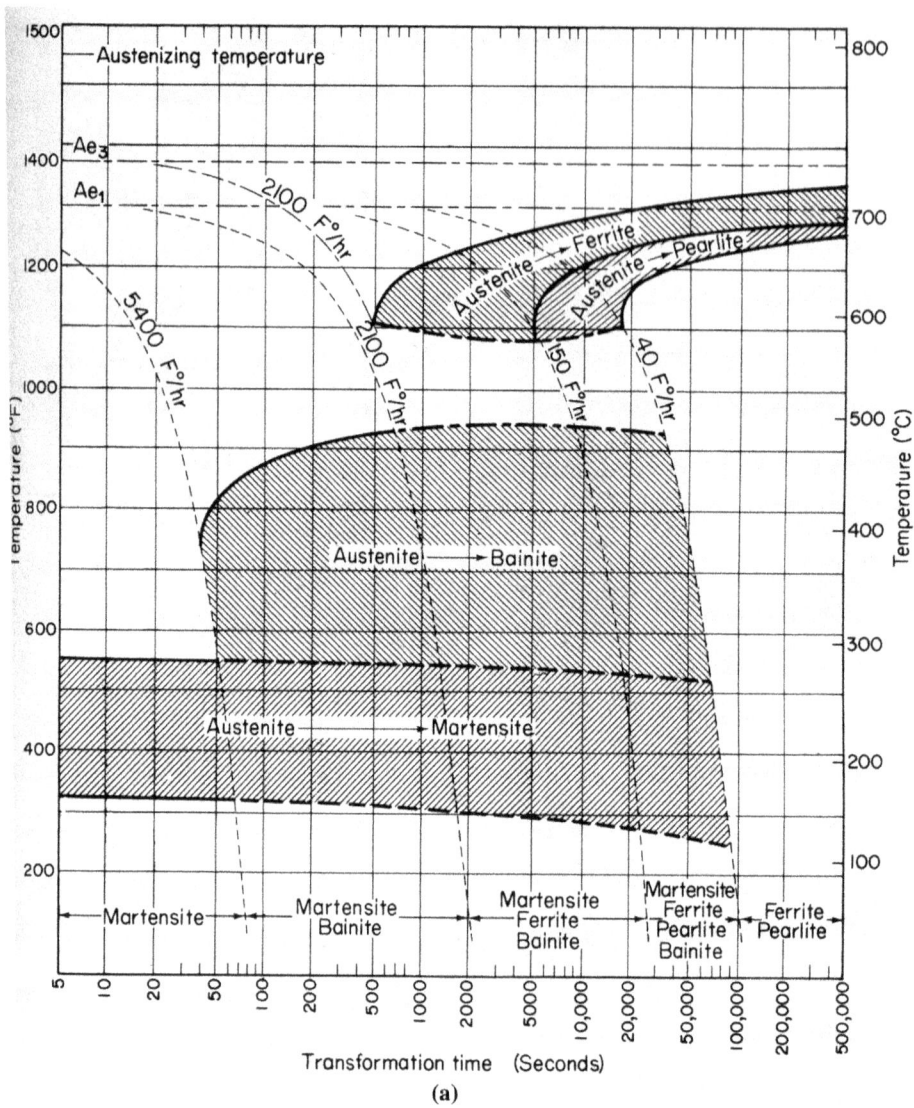

Figure 11 Pollack, Herman W. 1977. *Materials Science and Metallurgy*. Reston, VA: Reston Publishing Co., Inc. Fig. 8-6 (a), pg. 175.

Appendix VIII: Popular Knife-Makers' Steels, c. 1973

An early 21st century version of this 1973 excerpt from Latham's *Knives and Knifemakers* would probably contain significant alterations in the form of additional chromium-vanadium alloy combinations. An unresolved controversial question lingers: in contrast to the lack of significant improvement in steel for woodworking edge tools and plane blades, are late 20th century and early 21st century "chromium-vanadium" knife-makers' steels superior to 18th and 19th century products?

Composition of the most popular knifemaker's steels
This list shows the percentage of each element in each steel. Iron, of course, accounts for at least 80 percent of these steels; only the elements added to the iron are listed here.

154-Cm			W-2		
	Carbon	1.05		Carbon	.06/1.4
	Manganese	0.60		Manganese	.25
	Phosphorus	0.030		Silicon	.25
	Sulfur	0.030		Vanadium	.25
	Silicon	0.25			
	Chromium	14.00	M-2	Carbon	.85
	Molybdenum	4.00		Manganese	.25
				Phosphorus	.00 .03 max.
440-C	Carbon	1.00		Sulfur	.00 .03 max.
	Manganese	.50		Silicon	.30
	Silicon	.40		Chromium	4.20
	Chromium	17.05		Molybdenum	5.00
	Molybdenum	.45		Tungsten	6.35
	Nickel	.20		Vanadium	1.90
F-8	Carbon	1.30	A-2	Carbon	1.00
	Tungsten	8.00		Manganese	.50/.70
	Chromium	4.00		Silicon	.25/.40
	Vanadium	.25		Chromium	5.00
				Molybdenum	1.00
D-2	Carbon	1.50			
	Manganese	.25/.40	O-1	Carbon	0.90
	Silicon	.30/.50		Tungsten	.50
	Chromium	11.50		Manganese	1.35
	Molybdenum	1.00		Silicon	.35
	Vanadium	.90		Chromium	.50

Figure 12 Latham, Sid. 1973. *Knives and Knifemakers*. New York: MacMillan. pg. 33.

Appendix IX: Principal Bibliographic Sources

Many of the citations in this bibliography will be encountered scattered through the general bibliographies of the *Hand Tools in History* series. They are cited again here because they are quoted or referenced within the *Glossary*; they constitute the principal sources of information on ferrous metallurgy used by the Davistown Museum. Hopefully, readers of this glossary will find these citations as useful as we have, despite their obscurity or difficulty to procure through interlibrary loan services. The notation "IS." at the end of a citation indicates it is "in stock" in our library.

Please note: a second general bibliography on metallurgy follows the *Glossary* bibliography. Many important citations on metallurgy and the history of tools can be found in the bibliographies to Vol 6: *Steel- and Toolmaking Strategies and Techniques before 1870*; Vol 7: *Art of the Edge Tool: The Ferrous Metallurgy of New England Shipsmiths and Toolmakers 1607 - 1882*; and Vol 8: *The Classic Period of American Toolmaking, 1827-1930*. Please note that volume 8 also includes numerous special topic bibliographies, including US and New England Toolmakers, Tools of the Trades, and Collectors' Guides, Handbooks, and Dictionaries. The Tools of the Trades bibliography is further subdivided into 64 special topics. The entire collection of the Davistown Museum's bibliographic citations is published in hard copy as Special Publication 48: *Davistown Museum: The Complete Bibliographies*, which is available online at http://www.davistownmuseum.org/bibliography.html.

Abell, Sir Wescott. 1948. *The Shipwright's Trade*. Cambridge: The University Press. IS.

Angell, Marcia. June 23, 2011. The epidemic of mental illness: Why? *New York Review of Books*. http://www.nybooks.com/articles/archives/2011/jun/23/epidemic-mental-illness-why/?pagination=false.

Appleton. 1866. *Appleton's dictionary of machines, mechanics, engine-work, and engineering.* vol 1. New York: D. Appleton and Company.

Aston, James and Story, Edward B. 1939. *Wrought iron: Its manufacture, characteristics and applications*. Pittsburgh, PA: A. M. Byers Company. IS.

Audel, Theo. 1942. *Audels Mechanical Dictionary*. NY: Theo Audel and Company. IS.

Bain, Edgar C. 1939. *Functions of the alloying elements in steel*. Cleveland, OH: American Society for Metals. IS. http://www.msm.cam.ac.uk/phase-trans/2004/Bain.Alloying/ecbain.html

Baker, William A. 1973. *A maritime history of Bath, Maine and the Kennebec region*. 2 vols. Bath, ME: Maritime Research Society of Bath. IS.

Barrett, Charles S. 1943. *Structure of metals: Crystallographic methods, principles and data*. New York: McGraw-Hill. IS.

Barraclough, K. C. 1984a. *Steelmaking before Bessemer: Blister steel, the birth of an industry*. Vol 1. London: The Metals Society.

Barraclough, K. C. 1984b. *Steelmaking before Bessemer: Crucible steel, the growth of technology*. Vol 2. London: The Metals Society.

Barraclough, K. C. 1991. Steel in the Industrial Revolution. In *The Industrial Revolution in metals*. Day, J. and Tylecote, R. F. eds. London: The Institute of Metals.

Baumeister, Theodore, ed. 1958. *Mechanical engineers' handbook*. New York: McGraw-Hill Book Company.

Bealer, Alex W. 1976. *The art of blacksmithing*. Edison, NJ: Castle Books.

Bergman, Torbern. 1781. *Dissertatio chemica de analysi ferri*. Uppsala, Sweden.

Bining, Arthur Cecil. 1933. *British regulation of the colonial iron industry*. Philadelphia: University of Pennsylvania Press. IS.

Biringuccio, Vannoccio. [1540] 1990. *The pirotechnia of Vannoccio Biringuccio: The classic sixteenth-century treatise on metals and metallurgy*. Trans. and Ed. Cyril Stanley Smith and Martha Teach Gnudi. Mineola, NY: Dover. IS.

Biswas, A. K. 1994. Iron and steel in pre-modern India- a critical review. *Indian Journal of History of Science*. 29: 579-610.

Bronson, Bennet. 1986. The making and selling of Wootz, a crucible steel of India. *Archaeomaterials*. 1:13-51.

Brack, H. G. 2005. *What needs to be retrieved: The marriage of tools, art and history*. Hulls Cove, ME: Pennywheel Press. IS.

Brack, H. G. 2006. *Norumbega reconsidered: Mawooshen and the Wawenoc diaspora*. Hulls Cove, ME: Pennywheel Press. IS.

Brack, H. G. 2008. *Art of the edge tool: The ferrous metallurgy of New England shipsmiths and edge toolmakers*. vol. 8. Hand Tools in History Series. Hulls Cove, ME: Pennywheel Press. IS.

Brack, H. G. 2010. *The phenomenology of tools*. West Jonesport, ME: Pennywheel Press. IS.

Brack, H. G. 2013. *Tools teach: An iconography of American hand tools*. Hulls Cove, ME: Pennywheel Press. IS.

Brain, Jeffery Phipps. 2007. *Fort St. George archaeological investigation of the 1607-08 Popham Colony*. Occasional Publications of the Maine Archaeological Society. Augusta,

ME: Maine State Museum and Maine Historical Preservation Commission. IS.

Brearley, H. 1933. *Steel makers*. London: Longmans Green.

Breant, J. R. 1823. Description d'un procede al'aide duquel on obtient une espece d'acier fondu, sembable a celui des lames damassees Orientales. *Bull. Societe d'Encouragement pour l'Industrie Nationale.* 22: 222-227.

Brewington, M. V. 1962. Shipcarvers of North America. Barre: Barre Publishing Company.

Bronson, B. 1986. The making and selling of Wootz, a crucible steel of India. *Archaeomaterials*. 1: 13-51.

Buchanan, F. 1807. *A journey from Madras through the countries of Mysore, Canara and Malabar*. 3 vols. London.

Camp, J. M. and Francis, C.B. 1919. *The making, shaping, and treating of steel*. Pittsburgh, PA: U.S. Steel Corp.

Cias, Witold W. n.d. *Phase transformation kinetics and hardenability of medium-carbon alloy steels*. Greenwich, CT: Climax Molybdenum Company. IS.

Cleere, Henry and Crossley, David. 1985. *The iron industry of the Weald*. Avon: Leicester University Press. IS.

Climax Molybdenum Co. 1967. *Symposium: A transformation and hardenability of steels: February 27-28th*. Sponsored by Climax Molybdenum Company of Michigan and the University of Michigan, Department of Chemical and Metallurgical Engineering, Ann Arbor, MI.

Craddock, P. T. 1995. *Early metal mining and production*. Edinburgh, Scotland: Edinburgh University Press.

Craddock, P. T. 1998. New light on the production of crucible steel in Asia. *Bulletin of the Metals Museum of the Japan Institute of Metals*. 29: 41-66.

Craddock, P. T. and Hughes, M. J. eds. 1985. *Furnaces and smelting technology in antiquity*. British Museum Occasional Paper 48. London. IS.

Craddock, P. T. and Lang, eds. 2003. *Mining and metal production through the ages*. London: British Museum Press.

Dane, E. Surrey. 1973. *Peter Stubs and the Lancashire hand tool industry*. Altrincham: John Sherratt and Son Lt. IS.

Davenport, E. S. and Bain, E. C. 1930. *TRANS. AIME (The American Institute of Mining, Metallurgical and Petroleum Engineers)*. 117.

Day, Peter. 1995. Michael Faraday as materials scientist, history of materials. *Materials*

World.

Egerton, W. 1896. *Indian and Oriental armour*. London.

Faraday, M. 1819. An analysis of wootz or Indian steel. *Quarterly Journal of Science, Literature, and the Arts*. 7: 319-30.

Fisher, Douglas A. 1949. *Steelmaking in America*. New York: US Steel Corp.

Fisher, Douglas Alan. 1963. *The epic of steel*. New York: Harper & Row. IS.

Fostini, R. V. and Schoen, F. J. 1967. Affects of carbon and austenitic grain size on the hardenability of molybdenum steels. Paper presented at the Symposium: A transformation and hardenability of steels, February 27-28th in Ann Arbor, MI. 195-210.

Gardner, J. Starkie. 1892. *Victoria & Albert Museum: Ironwork: Part I. From the earliest times to the end of the mediaeval period*. London: Printed under the authority of the Board of Education.

Ghose, B. N., Bhattacharya, J., Das, N. K., De, R. K., Krishnan, C. S. S. R. and Mohanty, O. N. 1997. *Superplasticity in iron-carbon alloys*. Paper presented for ICSAM-97. Bangalore, India: Indian Institute of Science.

Goodman, W. L. 1964. *The history of woodworking tools*. New York: David McKay Company. IS.

Gordon, Robert. 1996. *American iron, 1607 - 1900*. Baltimore: Johns Hopkins University Press. IS.

Gore, Albert. 2006. *An inconvenient truth*. Movie. Paramount Classics.

Greener, W. W. 1910. *The gun and its development*. New York: Bonanza Books. IS.

Grossman, M. A. 1935. *The principles of heat treatment*. Cleveland, OH: American Society for Metals.

Habraken, L. J. and Economopoulous, M. 1967. Bainitic microstructures in low-carbon alloy steels and their mechanical properties. *Papers presented at the Symposium: A transformation and hardenability of steels, February 27-28th in Ann Arbor, MI*. 69-108.

Hayman, Richard and Horton, Wendy. 2003. *Ironbridge: History & guide*. Stroud: Tempus Publishing Ltd.

Heavrin, Charles A. 1998. *The axe and man: The history of man's early technology as exemplified by his axe*. Mendham, NJ: The Astragal Press. IS.

Howe, Henry M. 1890. *The metallurgy of steel*. New York: Scientific Publishing Co.

Hunter, Louis C. 1979. *Waterpower in the century of the steam engine*. Vol. 1 of *A history of industrial power in the United States, 1780-1930*. Published for the Eleutherian Mills-Hagley Foundation. Charlottesville, VA: University Press of Virginia.

Hunter, Louis C. 1985. *Steam power.* Vol. 2 of *A history of industrial power in the United States, 1780-1930*. Published for the Eleutherian Mills-Hagley Foundation. Charlottesville, VA: University Press of Virginia.

Hunter, Louis C. 1991. *The transmission of power.* Vol. 3 of *A history of industrial power in the United States, 1780-1930*. Cambridge, MA: The MIT Press.

Illich, Ivan. 1973. *Tools for conviviality*. New York: Harper & Row. IS.

Kauffman, Henry J. 1972. *American axes: A survey of their development and their makers.* Brattleboro, VT: The Stephen Greene Press. IS.

Kelly, Jack. 2004. *Gunpowder: Alchemy, bombards, and pyrotechnics: The history of the explosive that changed the world*. NY: Basic Books.

Kinsman, K. R. and Aaronson, H. I. 1967. Influence of molybdenum and manganese on the kinetics of the proeutectoid ferrite reaction. *Papers presented at the Symposium: A transformation and hardenability of steels, February 27-28th in Ann Arbor, MI*. 39-56.

Klenman, Allen. 1990. *Axe makers of North America*. Victoria, BC: Whistle Punk Books. IS.

Knight, Edward H. 1877. *American mechanical dictionary*. Cambridge: Cambridge University Press. IS.

Kolchin, B.A. 1953. *Chernoe Metallurgie v Drevnoi Rusi.* Moscow.

La Nice, Susan, Hook, Duncan and Kraddock, Paul. eds. 2007. *Metals and mining studies in Archaeometallurgy*. London: Archtype Publications in association with the British Museum.

Latham, Sid. 1973. *Knives and knifemakers*. New York: MacMillan.

Lee, Leonard. 1995. *The complete guide to sharpening*. Newtown, CT: The Taunton Press.

Levin, A. A., Meyer, D. C., Reibold, M., Kochmann, W., Patzke, N. and Paufler, P. 2005. Microstructure of a genuine Damascus sabre. *Cryst. Res. Technol.* 40: 905-16.

Light, John D. 2007. A dictionary of blacksmithing terms. *Historical Archaeology*. 41: 84-157.

Liu, K. H., Chan, H., Notis, M. R. and Pigott ,V. C. 1984. Analytical electron microscopy of early steel from the Bacqah Valley, Jordan. In *Microbeam analysis.* Romig, A. D. Jr. and Goldstein, J. I. eds. San Francisco: San Francisco Press, Inc.

Lowe, T. L. 1989. Solidification and the crucible processing of Deccani ancient steel. In *Principles of solidification and materials processing*, vol. 2, Trivedi, R., Sekhar, J. A. and Mazumdar, J. eds. New Delhi: Oxford and IBH Publishing.

Lowe, T. L. 1990. Refractories in high-carbon iron processing: A preliminary study of

Deccani Wootz-making crucibles. In *Ceramics and civilization.* Kingery, W. D. ed. Pittsburgh, PA: The American Ceramic Society.

Lytle, Thomas G. 1984. *Harpoons and other whalecraft.* New Bedford, MA: The Old Dartmouth Historical Society Whaling Museum.

Lynch, Kenneth. 2007. *The armourer and his tools: The Kenneth Lynch tool collection.* Davistown Museum Special Publication 45. Hulls Cove, ME: Pennywheel Press. IS.

Maiklen, Lara, ed. 1998. *Ultimate visual dictionary of science.* New York: DK Publishing, Inc.

Mehl, R. F. 1938. *Symposium on the hardenability of alloy steels.* Cleveland, OH: American Society for Metals.

Mehl, Robert F. and Hagel, W. C. 1956. *The austenite-pearlite reaction, progress in metal physics.* vol. 6. Oxford: Pergamon Press.

Mehl, Robert F. and Wells, Cyril. 1937. *Constitution of high purity iron-carbon alloys.* Technical Publication No. 798. American Institute of Mining and Metallurgical Engineers.

Mercer, Henry C. [1929] 1975. *Ancient carpenters' tools.* Horizon Press for the Bucks County Historical Society. IS.

Moxon, Joseph. [1703] 1989. *Mechanick exercises or the doctrine of handiworks.* Morristown, NJ: The Astragal Press.

- The first issue of *The Doctrine of Handiworks* was published in London in 1677, and remains the definitive late Renaissance commentary on blacksmithing and the sources of steel available in the late 17th century.

Mushet, D. 1804. Experiments on wootz or Indian steel. *Philosophical Transactions of the Royal Society.* 95:175.

Needham, J. 1958. *The development of iron and steel technology in China.* 2 Vols. London: Newcomen Society.

- The definitive survey of steelmaking techniques in China before the modern era.

Nieh, T. G., Wadsworth, J. and Sherby, O. D. 2005. *Superplasticity in metals and ceramics.* Cambridge Solid State Science Series. Cambridge: Cambridge University Press.

Nizolek, T., Benscoter, A. and Notis, M. R. 2007. Metallography and microanalysis of pattern-welded composite steels. *Microsc. Microanal.* 13(Suppl 2).

NUCOR. 2007. *Certified test report.* Auburn, NY: NUCOR Steel Auburn, Inc.

Oberg, Erik, Jones, Franklin D., Horton, Holbrook L. and Ryffel, Henry H. 2000. in

Machinery's handbook. McCauley, Christopher J., Heald, Riccardo and Hussain, Muhammed Iqbal eds. New York: Industrial Press Inc.

Oblak, J. M. and Hehemann, R. F. 1967. The structure and growth of Widmanstätten ferrite. *Papers presented at the Symposium: A transformation and hardenability of steels, February 27-28th in Ann Arbor, MI*. 15-38.

Oxford English Dictionary (OED). 1975. Oxford: Oxford University Press.

Palmer, Frank R. 1937. *Tool steel simplified: A handbook of modern practice for the man who makes tools*. Reading, PA: The Carpenter Steel Company. IS.

Partington, J. R. 1961. *A history of chemistry.* New York: St. Martin's Press.

Paxton, H. W. 1967. The formation of austenite. *Papers presented at the Symposium: A transformation and hardenability of steels, February 27-28th in Ann Arbor, MI*. 3-14.

Paynter, Sarah. 2007. Innovations in bloomery smelting in Iron Age and Romano-British England. In *Metals and mining studies in Archaeometallurgy*. La Nice, Susan, Hook, Duncan and Kraddock, Paul eds. 202-10. London: Archtype Publications in association with the British Museum.

Pense, Alan W. 2000. Iron through the ages. *Materials Characterization*. 45:353-63.

Perttula, J. 2004. Wootz Damascus steel of ancient Orient. *Scandinavian Journal of Metallurgy*. 33:92-7.

Petty, E. R. 1970. *Martensite*. New York: Longmans.

Piaskowski, Jerzy. 1978. Metallographic examination of two damascene steel blades. *Journal for the History of Arabic Science*. 2: 3-30.

Piaskowski, Jerzy. 1982. On the manufacture of high-nickel iron Chalibean steel in antiquity. In *Early Pyrotechnology*. Wertime, Theodore ed. Washington: Smithsonian Institution Press. IS.

Pickering, F. B. 1967. The structures and properties of bainite in steels. *Papers presented at the Symposium: A transformation and hardenability of steels, February 27-28th in Ann Arbor, MI*. 109-132.

Pleiner, Radomir. 1969. Experimental smelting of steel in early medieval furnaces. *Pamatky archaeologicke*. 458-487.

Pleiner, Radomir. 1980. Early iron metallurgy in Europe. In *The coming of the age of iron*. Wertime, Theodore A. and Muhly, James D. eds. New Haven: Yale University Press.

Pollack, Herman W. 1977. *Materials science and metallurgy*. Reston, VA: Reston Publishing Co., Inc. IS.

Rao, K. N. P. 1989. Wootz-Indian crucible steel, feature article. No. 1. *Metal News*. 11:

1-6.

Rao, K. N. P., Mukherjee, J. K. and Lahiri, A. K. 1970. Some observations on the structure of ancient steel from south India and its mode of production. *Bulletin of Historical Metallurgy*. 4: 12-4.

de Réaumur, René Antoine Ferchault. 1722. *L'art de convertir le fer forgé en acier*. Paris, France.

Renfrew, Colin. 1973. *Before civilization: The radiocarbon revolution and prehistoric Europe*. London: Cambridge University Press. IS.

Russell, Carl P. 1967. *Firearms, traps and tools of the mountain men*. New York: Knopf.

Salaman, R. A. 1975. *Dictionary of tools used in the woodworking trades, c. 1700-1975*. New York: Charles Scribner's Sons.

Schmitt, Catherine, Belliveau, Mike, Donahue, Rick and Sears, Amanda. 2007. *Body of evidence – a study of pollution in Maine people*. Portland, ME: Alliance for a Clean and Healthy Maine. IS.

Schubert, John Rudolph Theodore. 1957. *The history of the British iron and steel industry from c. 450 B.C. to A.D. 1775*. London: Routledge & Kegan Paul.

Sellens, Alvin. 1990. *Dictionary of American hand tools: A pictorial synopsis*. Augusta, KS: Alvin Sellens.

Sherby, Oleg D. 1995a. Damascus steel and superplasticity – Part I: Background, superplasticity, and genuine Damascus steels. *SAMPE Journal*. 31:10-7.

Sherby, Oleg D. 1995b. Damascus steel and superplasticity – Part II: Welded Damascus steels. *SAMPE Journal*. 31.

Shrager, M. 1961. *Elementary metallurgy and metallography*. New York: Dover. IS.

Sinha, Anil Kumar. 2003. *Physical metallurgy handbook*. New York: McGraw-Hill.

Sisco, Anneliese G. trans. 1956. *Réaumur's memoirs on steel and iron*. Chicago: University of Chicago Press.

Smith, Cyril S. 1960. *A history of metallography: The development of ideas on the structure of metals before 1890*. Chicago: University of Chicago Press. IS.

Smith, Cyril Stanley, ed. 1968. *Sources for the history of the science of steel 1532 - 1786*. Cambridge: The Society for the History of Technology and MIT Press. IS.

Smith, Cyril Stanley. 1983. *A search for structure: Selected essays on science, art and history*. Cambridge: MIT Press.

Society of Automotive Engineers, Inc., and American Society for Testing and Materials. 1986. *Metals & alloys in the unified numbering system with a description of the system and a cross index of chemically similar specifications.* Warrendale, PA: Society of Automotive Engineers, Inc.

Spring, Laverne W. 1992. *Non-technical chats on iron and steel and their application to modern industry.* Bradley, IL: Lindsay Publications Inc. IS.

Srinivasan, S. and Griffiths, D. 1997. South Indian wootz: evidence for high-carbon steel from crucibles from a newly identified site and preliminary comparisons with related finds. *Material Issues in Art and Archaeology-V, Materials Research Society Symposium Proceedings Series.* Vol. 462. Pittsburgh.

Srinivasan, Sharada, and Ranganathan, Srinivasa. n.d. *India's legendary 'Wootz' steel: An advanced material of the ancient world.* Bangalore, India: National Institute of Advanced Studies and Indian Institute of Science. http://materials.iisc.ernet.in/~wootz/heritage/WOOTZ.htm

Stanley, Philip E. 1984. *Boxwood & ivory: Stanley traditional rules, 1855 - 1975.* Westborough, MA: The Stanley Publishing Co. IS.

Stanley Rule & Level Co. 1975. *1859 Price list of boxwood and ivory rules, levels, try squares, sliding T bevels, gauges, &c., manufactured by the Stanley Rule and Level Company, also including the price list of boxwood and ivory rules manufactured by A. Stanley & Co., New Britain, Conn. Jan. 1855.* Fitzwilliam, NH: Ken Roberts Publishing Co.

Stodart, J. 1818. A brief account of Wootz. *Asiatic Journal.* 5.

Stodart, J. and Faraday, M. 1822. On the alloys of steel. *Philosophical Transactions of the Royal Society of London.* 112: 253-70.

Stone, George Cameron. 1999. *A glossary of the construction, decoration, and use of arms and armor in all countries in all times.* Mineola, NY: Dover Publications.

Story, Dana. 1995. *The shipbuilders of Essex: A chronicle of Yankee endeavor.* Gloucester, MA: Ten Pound Island Book Company.

Tomas, Estanislau. 1999. The Catalan process for the direct production of malleable iron and its spread to Europe and the Americas. *Contributions to Science.* 1:225-32.

Tresemer, David. 1996. *The scythe book: Mowing hay, cutting weeds, and harvesting small grains with hand tools.* Chambersburg, PA: Alan C. Hood & Co., Inc.

Tweedale, Geoffrey. 1987. *Sheffield steel and America: A century of commercial and technological interdependence, 1830-1930.* New York: Cambridge University Press. IS.

Tylecote, Ronald F. 1987. *The early history of metallurgy in Europe.* London: Longmans Green. IS.

Verhoeven, John D. 1987. Damascus steel, Part I: Indian Wootz steel. *Metallography.* 20:

145-51.

Verhoeven, John D. 2001. The mystery of Damascus blades. *Scientific American Magazine*. 284: 74-80.

Verhoeven, John D., Baker, H. H., Peterson, D. T., Clark, H. F. and Yater, W. M. 1990. Damascus steel, part III: The Wadsworth-Sherby mechanism. *Materials Characterization*. 24: 205-27.

Verhoeven, John D., Pendray, A. H. and Gibson, E. D. Wootz Damascus steel blades. *Materials Characterization*. 37: 9-22.

Wadsworth, J. and Sherby, O. D. 1978. Influence of chromium on superplasticity in ultra-high carbon steels. *Journal of Materials Science*. 13: 2645-49.

Wadsworth, J. and Sherby, O. D. 1980. On the Bulat-Damascus steels revisited. *Progress in Materials Science*. 25: 35-67.

Wayman, Michael L. 2000. *The ferrous metallurgy of early clocks and watches: Studies in post medieval steel*. Occasional Paper Number 136. London: British Museum. IS.

Wertime, Theodore A. 1962. *The coming of the age of steel*. Chicago: University of Chicago Press. IS.

Wertime, Theodore A. 1982. *Early pyrotechnology*. Washington: Smithsonian Institution Press.

Wertime, Theodore A. and Muhly, James D., eds. 1980. *The coming of the age of iron*. New Haven: Yale University Press. IS.

Weston, Thomas. 1906. *History of the town of Middleboro Massachusetts*. New York: Houghton, Mifflin and Company.

Whitworth, Joseph and Wallis, George. 1854. *The industry of the United States in machinery, manufactures, and useful and ornamental arts compiled from the official reports of Messrs. Whitworth and Wallis*. London: George Routledge & Co.

Williams, Alan. 2007. Crucible steel in medieval swords. In *Metals and mining studies in Archaeometallurgy*. La Nice, Susan, Hook, Duncan and Kraddock, Paul eds. London: Archtype Publications in association with the British Museum.

Zaky, A. R. 1979. *Medieval Arab arms, Islamic arms and armour*. London.

Appendix X: Metallurgy and Metalworking Bibliography

The special focus of the Davistown Museum bibliographies in the *Hand Tools in History* series is ferrous metallurgy from the early Iron Age to the beginning of the modern era. Numerous references pertaining to modern iron- and steelmaking technologies are included in our bibliographies, which are nonetheless *not* intended to be a complete survey of important references on this topic.

Aitchison, Leslie. (1960). *A history of metals*. 2 volumes. Interscience Publishers, NY, NY.

Allen, Richard Sanders. (1992). Connecticut iron and steel from Black Sea sands. *Journal of the Society for Industrial Archeology*. 18(1/2). pg. 129-132.

Alling, George W. (1903). *Points for buyers and users of tool steel*. D. Williams, New York, N.Y.

Arnold, J. O. (1898). The micro-chemistry of cementation. *Journal of Iron and Steel Institute*. Part II. pg. 185-194.

Aston, James and Story, Edward B. (1939). *Wrought iron: Its manufacture, characteristics and applications*. A. M. Byers Company, Pittsburgh, PA. IS.

Bancroft, W. E. (1946). Salt baths for hardening high speed steel. *Metal Progress*. pg. 941-947.

Barraclough, K. C. (October 1971). Puddled steel: A forgotten chapter in the history of steelmaking. *J. Iron Steel Inst*. 209. pg. 785-789.

Barraclough, K. C. (December 1971). Puddled steel: The technology. *J. Iron Steel Inst*. 209. pg. 952-957.

Barraclough, K. C. (1974). The production of steel in Britain by the cementation and crucible processes. *Historical Metallurgy*. 8. pg. 103-111.

Barraclough, K. C. (1976). The development of the cementation process for the manufacture of steel. *Post-Medieval Archaeology*. 10. pg. 65-88.

Barraclough, K. C. (1978). A crucible steel melter's logbook. *Historical Metallurgy*. 12.

Barrett, Charles S. (1943). *Structure of metals: Crystallographic methods, principles and data*. McGraw-Hill, New York, NY.

Batra, R.C. (1994). *Consideration of microstructural changes in the study of adiabatic shear bands*. Research Triangle Park, NC: U.S. Army Research Park. X.

Bauerman, H. (1891). *Elements of metallurgy*. 3rd edition. Phillips, J.A., ed. London,

UK.

Baylis, B. (1866). *On puddling -- by a practical puddler*. Published privately, London, UK.

Bell, Isaac Lowthian. (1872). *Chemical phenomena of iron smelting: An experimental and practical examination of the circumstances which determine the capacity of the blast furnace, the temperature of the air, and the proper condition of the materials to be operated upon*. D. Van Nostrand, NY, NY.

Bell, Isaac Lowthian. (1878). On the separation of phosphorus from pig iron. *Journal of Iron and Steel Institute.* 1. pg. 17-34, 34-37.

Bell, Isaac Lowthian. (1884). *Principles of the metallurgy of iron and steel.* Newcastle, UK.

Bell Telephone System. *Principles of zone melting.* Monograph 2000.

Bell Telephone System. *Ultrapure metals produced by zone melting technique.* Monograph 2147.

Bell Telephone System. *Continuous multistage separation by zone melting.* Monograph 2388.

Bell Telephone System. *Single crystals of exceptional quality by zone leveling.* Monograph 2626.

Bergman, Torbern. (1781). *Dissertatio chemica de analysi ferri.* Uppsala, Sweden.

- Extracts from this work (*A Chemical Essay on the Analysis of Iron*) are reprinted in English in C. S. Smith's *Sources for the history of the science of steel 1532 - 1786.*

Bessemer, Henry, Sir. (August 13, 1856). The manufacture of iron (steel) without fuel. *Cheltenham Meeting of the British Association [of iron makers].*

- A complete copy of this paper may be found on page 156 of *Sir Henry Bessemer, F.R.S. an autobiography*. http://books.google.com/books?id=XOA-AAAAYAAJ&dq=bessemer+%22The+manufacture+of+malleable+iron+without+fuel%22&source=gbs_navlinks_s.

Bethlehem Steel Company. (1944). *Properties of frequently used carbon and alloy steels.* Bethlehem, PA.

Bethlehem Steel Company. (1944). *Steel in the making.* Bethlehem, PA.

Bethlehem Steel Company. (1942). *Tool steel treaters' guide.* Bethlehem, PA.

Bezis-Selfa, John. (2004). *Forging America: Ironworkers, adventurers and the Industrial Revolution*. Cornell University Press, Ithaca, NY.

Birch, Alan. (1952). Midlands iron industry during the Napoleonic wars. *Edgar Allen News*. 31. pg. 231-233.

Boucher, Jack E. (1964). *Of Batsto and bog iron.* The Batsto Citizens Committee, Batsto, NJ. IS.

Brearley, Harry. (February 2, 1924). Stainless steel: The story of its discovery. *Sheffield Daily Independent.*

Brearley, Harry. (1941). *Knotted string: An autobiography of a steel-maker.* Longmans.

Brearley, Harry. (1946). *Talks about steel making.* American Society for Metals, Cleveland, Ohio.

Brownlie, D. (1930). The history of the cementation process of steel manufacture. *Journal of Iron and Steel Institute.* CXXI, Part I. pg. 455-464, 474.

Bullens, D. K. (1938/1939). *Steel and its treatment.* 2 vols. John Wiley and Sons, Inc., New York, NY.

Camp, J. M. and Francis, C. B. (1919). *The making, shaping, and treating of steel.* Pittsburgh, PA.

Campbell, H. H. (1907). *The manufacture and properties of iron and steel.* Hill Publishing Co., NY, NY.

Campbell, H. H. (1896). *The manufacture and properties of structural steel.* Scientific Publishing Co., NY, NY.

Campbell, H. L. (date unknown). *Metal casings.* John Wiley and Sons, Inc., New York, NY.

Carlberg, Per. (July 1958). Early industrial production of Bessemer steel at Edsken. *Journal of the Iron and Steel Institute.* 189. pg. 201-204.

Carpenter, H. C. H. and Robertson, J. M. (1930). The metallography of some ancient Egyptian implements. *Journal of Iron and Steel Institute.* CXXI, Part I. pg. 417-448.

Carpenter Steel Company. (1944). *Carpenter matched tool steel manual.* Seventh edition. The Carpenter Steel Company, Reading, PA. IS.

Charles, J. A. (1975). Where is the tin? *Antiquity.* 49. pg. 19-24.

Christian, J. W. (1965). *Physical properties of martensite and bainite.* Iron and Steel Institute, London.

Clarke, E.B. (1914). Electric furnaces for steelmaking. *TAES.* 25. pg. 139-159.

Coghlan, Herbert H. (1956). *Notes on prehistoric and early iron in the Old World.* Occasional Papers on Technology. 8. Oxford University Press. pg. 134-165.

Colby, A. L. (1903). Nickel steel: Its properties and applications. *Proceedings of the*

American Society of Testing Materials. 3. pg. 141-168.

Conner, William G. (1921-1922). The carbonizing of steel parts. In: *Transactions of American Society for Steel Treating.* Volume II. pg. 148.

Cottrell, Alan Howard. (1967). *An introduction to metallurgy.* Edward Arnold, London.

Craddock, Paul T. (date unknown). *The manufacture and properties of iron and steel.* McGraw-Hill Book Co., New York, NY.

Craddock, P. T., Ed. (1998). *2000 years of zinc and brass.* British Museum Occasional Paper 50, London, England.

Crossley, D. W. (1981). Medieval iron smelting. In: *Medieval industry.* Crossley, D.W. Ed. CBA Research Report. no. 40. Council for British Archaeology, London, UK. pg. 29-41.

Crookes, W. and Rohrig, E. A. (1870). *A practical treatise on metallurgy.* New York, NY.

Davis, James J. (1922). *The Iron puddler.* Bobbs-Merrill, Indianapolis, IN.

Dichmann, C. (n.d.). *The basic open-hearth steel process.* Constable & Co., Ltd., London.

Dixon, George F. (2004). *A blacksmith's craft.* Blue Moon Press, Huntingdon, PA.

Ducoff-Barone, Deborah. (1983). Marketing and manufacturing: A study of domestic cast iron articles produced at Colebrookdale furnace, Berks County, Pennsylvania, 1735-1751. *Pennsylvania History.* 50. pg. 20-37.

Elam, C. F. (1935). *The distortion of metal crystals.* Oxford.

Ellingham, H. J. T. (1944). The physical chemistry of process metallurgy. *Journal of the Society of Chemistry and Industry.* 63. pg. 125.

Espelund, Arne. (1997). The "Evenstad" process-description, excavation, experiment and metallurgical evaluation. In: Norbach, Lars Chr., ed., *Early iron production-archaeology, technology and experiments.* Technical Report Nr. 3. The Historical-Archaeological Experimental Centre, Lejre, Denmark. pg. 47-58.

Espelund, Arne. (1997). Ironmaking in Trondelag during the Roman and Pre-Roman Iron Age. In: Norbach, Lars Chr., ed., *Early iron production-archaeology, technology and experiments.* Technical Report Nr. 3. The Historical-Archaeological Experimental Centre, Lejre, Denmark. pg. 103-114.

Evenstad, O. (1968). A treatise on iron ore and the process of turning it into iron and steel. *Bull. HMG.* Translated by N.L. Jensen. 2(2). pg. 61-5.

Fairbairn, William. (1865). *Iron: Its history, properties and processes of manufacture.*

Adam and Charles Black, Edinburgh.

Fennimore, Donald L. (1996). *Metalwork in early America: Copper and its alloys from the Winterthur collection*. Henry Francis du Pont Winterthur Museum, Winterthur, DL.

Finley, M. I. (Sept 1970). Metals in the ancient world. *Journal of the Royal Society of Arts*. 4.

Flather, D. (1901-1902). Crucible steel: Its manufacture and treatment. *Proc. Staffs. Iron Steel Inst.*, pg. 55-82.

Forsythe, Robert. (1908). *The blast furnace and the manufacture of pig iron: An elementary treatise for the use of the metallurgical student and the furnaceman*. D. Williams Co., NY, NY.

Fruehan, R. J. (1998). *The making, shaping, and treating of steel*. 11th edition. Vol. 2 Steelmaking and refining volume. AISE Steel Foundation, Pittsburgh, PA.

Fulford, Michael, Sim, David, Doig, Alistair, and Painter, Jon. (2005). In defence of Rome: A metallographic investigation of Roman ferrous armour from northern Britain. *Journal of Archaeological Science*. 32. pg. 241-150. IS.

Fuller, John, Sr. (1894). *Art of coppersmithing: A practical treatise on working sheet copper into all forms*. David William, NY, NY. Reprinted in 1993 by Astragal Press, Mendham, NJ. IS.

Gale, Walter Keith Vernon. (1963/1964). Wrought iron: A valediction. *Transactions of the Newcomen Society*. 36. pg. 1-11.

Gale, Walter Keith Vernon. (1964/1965). The rolling of iron. *Transactions of the Newcomen Society*. 37. pg. 35-46.

Gale, Walter Keith Vernon. (1979). *The black country iron industry*. Metals Society, London, UK.

Gardner, John. (March 1970). Cast steel. *The Chronicle*. 23(1). pg. 6, 16. IS.

Gilmer, H. (1953). Birth of the American crucible steel industry. *Western Pennsylvania Historical Mag*. XXXVI. pg. 19-34.

Giolitti, Frederico. (1915). *The cementation of iron and steel*. Translated from the Italian by Richards, Joseph W. and Rouiller, Charles A. McGraw-Hill, New York, NY.

Gledhill, J. M. (1904). The development and use of high speed steel. *Journal of Iron and Steel Institute*. II. pg. 127-167.

Gledhill, J. M. (1904). High speed tool steel. *Engineering Review*. pg. 405-411.

Goodale, S. L. (1920). *Chronology of iron and steel*. Pittsburgh, PA.

Gordon, Robert B. (1983). Material evidence of the development of metalworking technology at the Collins Axe Factory. *IA: The Journal of the Society for Industrial Archeology*. 9. pg. 19-28.

Gordon, Robert B. (1988). Strength and structure of wrought iron. *Archeomaterials*. 2. pg. 109-137.

Gordon, Robert B. (2001). *A landscape transformed: The ironmaking district of Salisbury, Connecticut*. Oxford University, New York, NY.

Gordon, Robert B. and Raber, Michael S. (1984). An early American integrated steelworks. *IA: The Journal of the Society for Industrial Archeology*. 10(1). pg. 17-34.

Gray, R. D. (1979). *Alloys and automobiles*. Indianapolis.

Greenwood, William Henry. (1870). *A manual of metallurgy*. G. P. Putnam's sons, New York, NY.

Greenwood, William Henry. (1902). *Steel and iron: Comprising the practice and theory of the several methods pursued in their manufacture, and of their treatment in the rolling mills, the forge and the foundry*. Cassell and Co., Ltd., London.

Greenwood, William Henry. (1907). *Iron: Its sources, properties, and manufacture*. Revised and partially rewritten by A. Humboldt Sexton. David McKay, Publishers, Philadelphia, PA.

Grossmann, Marcus Aurelius. (1935). *Principles of heat treatment*. United States Steel Corp. The Haddon Craftsmen, Inc., Scranton, PA. IS.

Grossman, M. A. and Bain, E. C. (1931). *High-speed steel*. New York, NY.

Guillet, L. (1905). *Les aciers speciaux, 2 Vols*. Paris, FR.

Guthrie, R. I. L. and Stubbs, P. (1973). Kinetics of scrap melting in baths of molten pig iron. *Canadian Metallurgical Quarterly*. 12(4). pg. 465-473.

Hackney, W. (1874/1875). The manufacture of steel. *Proc. Inst. Civil Eng*. xlii. pg. 2-68.

Hadfield, R. A. (1888). Manganese steel. *Journal of the Iron and Steel Institute of Japan*. 33. pg. 41-77.

Hadfield, R. A. (1888). Manganese and its applications in metallurgy. *Proc. Inst. Civil Eng*. cxiii. pg. 1-59.

Hadfield, R. A. (1889). On manganese steel. *Journal of the Iron and Steel Institute*. II. pg. 41-47.

Hadfield, R. A. (1889). On alloys of iron and silicon. *Journal of the Iron and Steel Institute*. II. pg. 222-242.

Hadfield, R. A. (1892). Alloys of iron and chromium. *Journal of the Iron and Steel Institute*. II. pg. 49-131.

Hadfield, R. A. (1894). The early history of crucible steel. *Journal of the Iron and Steel Institute of Japan.* II. pg. 224-238.

Hadfield, R. A. (1898/1899). Alloys of iron and nickel. *Proc. Inst. Civil Eng.* CXXVIII. pg. 1-125.

Hadfield, Robert. (January 1912). On Sinhalese iron and steel of ancient origin. *Proceedings of the Royal Society of London. Series A, Containing Papers of a Mathematical and Physical Character.* 86(584). pg. 94-100. X.

Hadfield, Robert Abbott, Sir. (1925). *Metallurgy and its influence on modern progress, with a survey of education and research.* Chapman & Hall, London.

Hadfield, Robert A., Sir. (1931). *Faraday and his metallurgical researches with special reference to their bearing on the development of alloy steels.* Chapman & Hall, London.

Hall, J. H. (April 3, 1913). The Manufacture of crucible steel. *Iron Trades Review.* pg. 791-793.

Hall, J. H. (April 10, 1913). The Manufacture of crucible steel. *Iron Trades Review.* pg. 849-856.

Harbord, Frank William and Hall. (1918). *The metallurgy of steel.* C. Griffin & Co., London.

Hasluck, Paul N., ed. (1904). *Metalworking: A book of tools, materials, and processes for the handyman.* Cassell and Company Ltd., London. Reprinted in 1994 by Lindsay Publications.

Hatch, Charles E. Jr. and Gregory, Thurlow Gates. (1962) The first American blast furnace, 1619 - 1622. *Virginia Magazine of History and Biography.* 70. pg. 259 - 296.

Hatfield, William Herbert. (1926). Comparison of permanent magnet steels. *J. Sci. Instrum.* 3. pg. 234-235.

Hatfield, William Herbert. (1928). *The application of science to the steel industry.* Printed by Evangelical Publishing House, Cleveland.

Hatfield, William Herbert. (1928). *Cast iron in the light of recent research.* Charles Griffin & Co., London.

Hatfield, William Herbert. (1938). *Research in the iron and steel industry.* Institute of Chemistry of Great Britain and Ireland, London.

Hatfield, William Herbert. (1942). *Lecture on rust-, acid- & heat-resisting steels.* Institute of Chemistry of Great Britain and Ireland, London.

Hibbard, Henry Deming. (1916). *Manufacture and uses of alloy steels.* G.P.O., Washington.

Hiles, J. and Mott, R. A. (1944). The mode of combustion of coke. *Fuel.* 23(6). pg. 154-

171.

Howe, H. M. (1890). *The metallurgy of steel*. The Scientific Publishing Co, NY, NY.

Hubbard, Howard G., ed. (November, 1937). Metal alloys. *The Chronicle*. 2(2). pg. 13. IS.

- This article continues over several issues: 2(2), pg. 13; 2(3), pg. 20; (2) 4, pg. 30; 2(5), pg. 39; 2(7), pg. 52.

- The information on the alloys is extracted from: Mackenzie's *Five Thousand Receipts in All the Useful and Domestic Arts,* Kay's edition, Philadelphia, 1829.

- "*Bronze*. Melt in a clean crucible 7 pounds of pure copper: when fused, throw into it 3 pounds of tin. These metals will combine, forming bronze, which, from the exactness of the impression which it takes from a mould, has, in ancient and modern times, been generally used in the formation of busts, medals, and statues." pg. 30.

- "*Bell metal.* Melt together 6 parts of copper, and 2 of tin: These proportions are the most approved for bells throughout Europe, and in China. In the union of the two metals above mentioned, the combination is so complete, that the specific gravity of the alloy is greater than that of the two metals combined." pg. 30.

Huddleston, Jerry and Stutzenberger, Fred. (July 2006). Damascene: Part I. *Muzzle Blasts*. 67(11). pg. 61-64. X.

Huddleston, Jerry and Stutzenberger, Fred. (August 2006). Damascene: Part II. *Muzzle Blasts*. 67(12). pg. 61-64. X.

Hyde, Charles. (1977). *Technological change in the British iron industry, 1700-1870*. Princeton University Press, Princeton, NJ.

International Correspondence Schools. (1916). *Gauges, jigs, dies, tempering, heat treatment, blacksmithing and forging*. I.C.S. Reference Library, International Textbook Company, Scranton, NY.

Ishihara, Y. (1958). Progress of special steelmaking in Japan. *Symposium on Production of Alloy Steels (Jamshedpur)*. pg. 172-175.

Jeans, J. S. (1880). *Steel: Its history, manufacture, properties and uses*. London.

Jefferson, T. B. and Woods, Gorham. (1954). *Metals and how to weld them*. The James F. Lincoln Arc Welding Foundation, Cleveland, OH. IS.

Johannsen, Otto. (1924). *Geschichte des eisens; im auftrage des Vereins deutscher eisenhüttenleute gemeinverstñdlich dargestellt, von dr. Otto Johannsen*. Verlag Stahleisen m. b. h., Düsseldorf.

- History of iron.

Kashchieva, Elena, Tsaneva, Svetila, Dimitriev, Yanko and Kirov, Roumen. (2003). Microstructure and chemical composition of Thracian medallions with enamel in millefiori technique found in Bulgaria. *Journal of Non-Crystalline Solids*. 323: 137-142.

Kauffman, Henry J. (n.d.). *Early American copper, tin and brass: Handcrafted metalware from colonial times*. Astragal Press, Mendham, NJ.

Kauffman, Henry J. (1948). *Coppersmithing in Pennsylvania: Being a treatise on the art of the eighteenth century coppersmith, together with a description of his products and his establishments*. Pennsylvania German Folk Lore Society, Allentown, PA.

Kauffman, Henry J. (1966). *Early American ironware: Cast and wrought*. Weathervane Books, NY, NY.

Kauffman, Henry J. (1966). *Metalworking trades in early America*. The Charles E. Tuttle Co., Inc. Reprinted in 1995 by Astragal Press, Mendham, NJ. IS.

Kauffman, Henry J. (1968). *American Copper & Brass*. T. Nelson, Camden, NJ. Reprinted in 1979 by Bonanza Books, NY, NY. IS.

Kauffman, Henry J. (December 1969). Cast steel. *The Chronicle*. 22(4). pg. 49-50. IS.

Kauffman, Henry J. (1970). *The American pewterer: His techniques & his products*. T. Nelson, Camden, NJ.

Kayser, J. F. (March 11, 1927). Practical crucible steel making. *Iron and Coal Trades Review*. pg. 396-397.

Kayser, J. F. (March 18, 1927). Practical crucible steel making. *Iron and Coal Trades Review*. pg. 438-439.

Kemp, Emory L. (1993). The introduction of cast and wrought iron in bridge building. *IA: The Journal of the Society for Industrial Archeology*. 19(2), pg. 5-16.

Kester Solder Company. (1948). *Solder and soldering technique: A condensed industrial manual*. Kester Solder Company, Chicago, IL. IS.

Killick, D. J. and Gordon, R. B. (1987). Microstructures of puddling slags from Fontley, England and Roxbury, Connecticut, U.S.A. *Journal of Historical Metallography Society*. 21. pg. 28-36.

Landrin, H. C. (1868). *Treatise on steel: Comprising its theory, metallurgy, properties, practical working, and use*. Translation of original French done by A. A. Fesquet. Henry Carey Baird, Philadelphia, PA.

Lankford, William T., Jr., Samways, N. L., Craven, R. F., McGannon, H. E., eds. (1985). *The making, shaping and treating of steel, 10th ed.* United States Steel Co., Herbick & Held, Pittsburgh, PA

Laughlin, Ledlie I. (1940). *Pewter in America*. 2 vols. Houghton Mifflin, Boston, MA.

Leever, Sylvia. (October 2003). Late medieval iron making, the bloomery process: A treatise on the making of iron in a bloomery and on the quality of the armour made from such iron. *TUDelft: Delft, Materials Science and Engineering.* Delft University of Technology, Delft, NL. IS.

Light, John D. (1987). Blacksmithing technology and forge construction. *Technology and Culture.* 28(3). pg. 658-665.

Light, John D., and Wylie, William N. (1986). A guide to research in the history of blacksmithing. *Research Bulletin.* No. 243, Parks Canada, Ottawa, Canada.

Lister, Raymond. (1966). *The craftsman in metal.* A. S. Barnes and Company, NY, NY. IS.

Longmuir, P. (1905). *Practical metallurgy: Iron and steel.* London.

Louis, H. (1879). On the chemistry of puddling. *J. Iron Steel Inst.* 1. pg. 219-226.

Lynch, John. (1957). *Metal sculpture: New forms new techniques.* The Viking Press, NY, NY. IS.

Markham, Eward Russell. (1903). *The American steel worker.* The Derry-Collard Company, NY, NY.

Martinez, M. A., Abenojar, J., Mota J. M. and Calabres, R. (2006). Ultra high carbon steels obtained by powder metallurgy. *Materials Science Forum.* 530-531: 328.33.

Maryon, Herbert. (1961). Early near eastern steel swords. *AJA.* 65. pg. 173-84.

Masse, H. J. L. J. (1911). *Chats on old pewter.* T. Fisher Unwin, London, UK.

McCreight, Tim and Bsullak, Nicole. (2001). *Color on metal: 50 artist share insights and techniques.* Guild Publishing, Shamper's Bluff, New Brunswick, Canada.

McCreight, Tim and Kazan, Katie. (2007). *Metalsmith's book of boxes and lockets.* Brynmorgan Press, Brunswick, ME.

Mehl, Robert Franklin. (1948). *A brief history of the science of metals.* Sponsored jointly by the Institute of Metals Division, AIME and the Seeley W. Mudd Memorial Fund, American Institute of Mining and Metallurgical Engineers, NY, NY.

Mehl, R. F. (1965). On the Widmannstatten structure. In: *The Sorby centennial symposium on the History of Metallurgy.* Smith, C. S., Gordon and Breach Science Publishers, NY.

Meilach, Dona and Seiden, Don. (1966). *Direct metal sculpture: Creative techniques and appreciation.* Crown Publishers Inc., NY, NY. IS.

Metal Powder Industries Federation. (2007). *Advances in powder metallurgy and particulate materials: Proceedings of the 2007 International Conference on Powder Metallurgy & Particulate Materials, May 13-16, Denver, Colorado.* Princeton, NJ:

MPIF.

Metallurgical Society. (1961). *History of iron and steelmaking in the United States: Publication in one book of a series of historical articles that have appeared in Journal of Metals, 1956-1961.* American Institute of Mining, Metallurgical, and Petroleum Engineers, NY, NY.

Misa, Thomas J. (1995). *A nation of steel*. London.

Moldenke, Richard George Gottlob. (1910). *The production of malleable castings: A practical treatise on the processes involved in the manufacture of malleable cast iron.* Penton Pub. Co., Cleveland, OH.

Moldenke, Richard George Gottlob. (1917). *The principles of iron founding*. McGraw-Hill Book Co., NY, NY.

Moldenke, Richard George Gottlob. (1927). *Malleable casting*. International Textbook Co., Scranton, PA.

Morton, G. R., and Brit, R. G. (1974). The Present day production of wrought iron. *Journal of the Historical Metallurgy Society.* 8. pg. 96-102.

Nosek, Elzbieta M. (1985). The Polish smelting experiments in furnaces with slag pits. In: Craddock, P.T. and Hughes, M.J., Eds. Furnaces and smelting technology in antiquity. British Museum Occasional Paper 48, London, England. IS.

Neilson, William G. (July 24, 1867). Manufacture of wrought iron direct from the ore. *Bulletin of the Iron and Steel Association.* 26.

Notis, M. and Shugar, A.(2003). Roman shears: Metallography, composition, and a historical approach to Investigation. *Proceedings of the Archaeometallurgy in Europe Conference, Milan, Italy Sept 24-26 2003.* 1. pg. 109-118. X.

Nuwer, Michael. (October 1988). From batch to flow: Production technology and work-force skills in the steel industry, 1880-1920. *Technology and Culture*. 29. pg. 808 - 838.

Oberg, Erik, Jones, Franklin D., Horton, Holbrook L. and Ryffel, Henry H. ([1914] 1996). *Machinery's Handbook: 25th edition: A reference book for the mechanical engineer, designer, manufacturing engineer, draftsman, toolmaker, and machinist.* Industrial Press Inc., NY, NY. IS.

Osborn, Henry Stafford. (1869). *The metallurgy of iron and steel: Theoretical and practical: In all its branches; with special reference to American materials and processes.* H.C. Baird, Philadelphia, PA.

Overman, Frederick. (1851). The manufacture of steel: Containing the practice and principles of working and making steel. A. Hart, Philadelphia, PA

Overman, Frederick. (1852). *A treatise on metallurgy; comprising mining, and general and particular metallurgical operations, with a description of charcoal, coke, and anthracite furnaces, blast machines, hot blast, forge hammers, rolling mills, etc., etc.* D. Appleton, NY, NY.

Overman, Frederick. (1854). *The manufacture of iron, in all its various branches. Including a description of wood-cutting, coal digging, and the burning of charcoal and coke; the digging and roasting of iron ore, the building and management of blast furnaces, working by charcoal, coke, or anthracite; the refining of iron. Also a description of forge hammers, rolling mills, blast machines, hot blast, etc. etc. To which is added an essay on the manufacture of steel.* H. C. Baird, Philadelphia, PA.

Overman, Frederick. (1894). *The manufacture of steel: Containing the practice and principles of working and making steel.* H.C. Baird, Philadelphia, PA.

Palmer, Frank R. (1937). *Tool steel simplified: A handbook of modern practice for the man who makes tools.* The Carpenter Steel Company, Reading, PA. IS.

Palmer, Frank R., Luerssen, George V. and Pendleton, Joseph S., Jr. (1978). *Tool steel simplified.* Fourth Edition. Chilton Company, Radnor, PA. IS.

Parr, J.G. (1972). The sinking of the Ma Robert: An excursion into mid nineteenth century steelmaking. *Technology and Culture.* 13(2). pg. 209-225.

Parsons, Sam Jones. (1909). *Malleable cast iron.* Van Nostrand, NY, NY.

Partington, J. R. (1961). *A history of chemistry.* St. Martin's Press, NY, NY.

Percy, J. (1864). *Metallurgy: Iron and steel.* London.

Percy, J. (1877). On the cause of the blisters on "Blister steel". *Journal of the Iron and Steel Institute.* II. pg. 460-463.

Perttula, Juha. 2001. Reproduced wootz Damascus steel. *Scandinavian Journal of Metallurgy.* 30:65-8.

Perttula, Juha. 2004. Wootz Damascus steel of ancient orient. *Scandinavian Journal of Metallurgy.* 33:92-7.

Phillips, J. Arthur. (1874). *Elements of metallurgy: A practical treatise on the art of extracting metals from their ores.* C. Griffin and Company, London, UK.

Piaskowski, J. (1972). Criteria for determining a technology of bloomery iron products. *Archaeologia Polona.* 17(1). pg. 7-45.

Pleiner, Radomir. (1969). *Iron working in ancient Greece.* National Technical Museum, Prague.

Pleiner, Radomir. (1980). Early iron metallurgy in Europe. In: *The Coming of the Age of Iron.* Wertime, Theodore A. and Muhly, James D., eds. Yale University Press, New Haven, CT. IS.

Piley, J. (1889). Alloys of nickel and steel. *Journal of the Iron and Steel Institute.* I. pg.

45-55.

Pollack, Herman W. (1977). *Materials science and metallurgy*. Reston Publishing Co., Inc., Reston, VA.

Pruemmer, R. A., Balakrishna Bhat, T., Siva Kumar, K. and Hokamoto, K. (2006). *Explosive compaction of powders and composites*. Enfield: Science Publishers.

Raquenet, A., Ed. (2010). *Artistry in iron: 183 designs, includes CD-ROM*. Dover Publications, NY.

Read, T. T. (1934). Metallurgical fallacies in archaeological literatur. *AJA*. 38. pg. 382.

Richardson, H. C. (1934). Iron, prehistoric and ancient. *AJA*. 38(4). pg. 574.

Ricketts, J. A. (2001). *History of Ironmaking*. Association for Iron & Steel Technology (AIST), Warrendale, PA.

Robb, Frances C. (1993). Cast aside: The first cast-iron bridge in the United States. *IA: The Journal of the Society for Industrial Archeology*. 19(2). pg. 48-62.

Roberts, G. A., Hamaker, J. C., and Johnson, A. R. (1962). *Tool steels*. American Society for Metals, Cleveland, OH.

Sabadasz, Joel. (1992). The development of modern blast furnace practice: The Monongahela Valley furnaces of the Carnegie Steel Company, 1872-1913. *IA: The Journal of the Society for Industrial Archeology*. 18(1/2). pg. 94-105.

Sauveur, A. (1910). *The metallography and heat treatment of iron and steel*. Cambridge, MA.

Schaur, R. Title unknown. *Stahl und Eisen*. 49. pg. 489-498.

Scoffern, J. et al. (1857). *The useful metals and their alloys.* London, UK.

Seely, Bruce E. (1981). Blast furnace technology in the mid-19th century: A case study of the Adirondack Iron and Steel Company. *IA: The Journal of the Society for Industrial Archeology*. 7(1). pg. 27-54.

Semiatin, S. L. and Lahoti, G. D. (1981). The forging of metals. *Scientific American*. 245(2). pg. 98-106.

Seymour, Lindsay, J. (1964). *Iron and brass implements of the English and American house*. Carl Jacobs, Bass River, MA.

Shrager, M. (1949). *Elementary Metallurgy and Metallography*. Reprinted in 1961 by Dover, NY. IS.

Siemens, C. W. (1862). On a regenerative gas furnace. *Proc. Inst. Mech. Eng.* pg. 21-44.

Siemens, C. W. (1868). On the regenerative gas furnace as applied to the manufacture of cast steel. *J. Chem. Soc.* pg. 279-310.

Siemens, C. W. (1873). On smelting iron and steel. *J. Chem. Soc.* pg. 661-678.

Slaager, Goud. (1978). *The gold beater, a miscellany.* Unpublished, prepared for the February 1978 meeting of the Early Trades and Crafts Society. IS.

Smith, C. S. (1964). The discovery of carbon in steel. *Technology and Culture.* 5(2). pg. 149-175.

Smith, C. S. (1965). *The Sorby centennial symposium on the history of metallurgy.* Gordon and Breach Science Publishers, NY.

Smith, J. B. (n.d.). *Wire manufacture and uses.* John Wiley & Sons, NY, NY.

Smith, Robert E. (1939). *Units in etching, spinning, raising, and tooling metal.* Edited by E. H. Mattingly. The McCormick-Mathers Publishing Company, Wichita, KS. IS.

Smith, Robert E. (1951). *Etching, spinning, raising and tooling metal.* McKnight & McKnight Publishing Company, Bloomington, IL. IS.

Snelus, G. J. (1879). On the removal of phosphorus and sulphur during the Bessemer and Siemens-Martin processes of steel manufacture. *Journal of the Iron and Steel Institute.* I. pg. 135-143.

Sorby, H. C. (1886). On the application of very high powers to the study of the microscopial structure of steel. *Journal of the Iron and Steel Institute of Japan.* I. pg. 140-147.

Sorby, H. C. (1887). The microscopical structures of iron and steel. *Journal of the Iron and Steel Institute of Japan.* 33. pg. 255-288.

Spring, Laverne W. (1917). *Non-technical chats on iron and steel and their application to modern industry.* Frederick A. Stokes Company, NY, NY. Reprinted in 1992 by Lindsay Publications Inc., Bradley, IL. IS.

- A surprisingly comprehensive overview of ferrous metallurgy and a museum favorite.
- Recommended for visitors to the Davistown Museum's Center for the Study of Early Tools library.
- Frequently cited in our *Glossary of Ferrous Metallurgy Terms.*

Stansfield, Alfred. (1907). *The electric furnace.* The Canadian Engineer, Toronto.

Steines, Adolph. (2002). *Moving metal: The art of chasing and repoussé.* Blue Moon Press, Huntingdon, PA.

- Translated from German.

Stodart, J. and Faraday, M. (1822). On the alloys of steel. *Philos. Trans. R. Soc.* 112. pg. 253-270.

Stoughton, B. (1923). *The metallurgy of iron and steel.* McGraw-Hill Book Co., NY, NY.

Straub, H., Tarmann, B. and Plöckinger, E. (1965). *Experiments on smelting in Noric-type furnaces.* Kantner Museums Schriften 35.

Swank, James M. (1892). *History of the manufacture of iron in all ages and particularly in the United States from colonial times to 1891.* Philadelphia, PA. IS.

Swank, James M. (1897). The manufacture of iron in New England. In: *The New England states.* Davis, William, ed. D.H. Hurd and Co., Boston, MA.

Tabor, D. (1951). *The hardness of metals.* Clarendon Press, Oxford, England.

Taylor, F. S. (1954). Some metallurgical processes of the early 16th century. *TNS.* 29. pg. 93-101.

Taylor, F. W. (1906). *On the art of cutting metals.* American Society of Mechanical Engineers, New York, NY.

Temple, Robert K. G. (1988). Cast iron and steel manufacture. *The UNESCO Courier.* 41(10).

Terekhova, N. N. (1974). Cast iron production technology of the Mongols in the Middle Ages. *Sov. Arch.* (1). pg. 69-78.

Thälen, L. (1973). Notes on the ancient iron currency bars of Northern Sweden and the nickel alloys of some archaeological objects. *Early Med. Stud.* 5. pg. 24-41.

Tholander, E. (1979). A study of the technology behind nickel-alloyed prehistoric steel. *Proc. 5th Atlantic Colloq., Dublin.* M. Ryan, Ed. pg. 317-34.

Tomtlund, J.-E. (1973). Metallographic investigation of 13 knives from Helgö. *Early Med. Stud.* 5. pg. 42-63.

Turner, Thomas Henry. (1895). *The metallurgy of iron and steel: Being one of a series of treatises on metallurgy written by associates of the Royal School of Mines.* Roberts-Austen, W.C. Ed. Charles Griffin, London, UK.

Turner, Thomas. (1900). *The metallurgy of iron.* C. Griffin and Co., London, UK.

Tweedale, Geoffrey. (1987). *Sheffield steel and America: A century of commercial and technological interdependence, 1830-1930.* Cambridge University Press, NY, NY. X.

Tweedale, Geoffrey. (1995). *Steel city: Entrepreneurship, strategy, and technology in Sheffield 1743-1993.* Oxford University Press, Oxford, UK.

Tylecote, R. F. (1965). Iron smelting in pre-industrial communities. *JISI.* 203. pg. 340-348.

Tylecote, R. F. (1981). Medieval smith and his methods. In: *Medieval industry.* D.W. Crossley, ed. CBA Research Report, No. 40. Council for British Archaeology, London,

England.

Tylecote, R. F., Austin, J. N. and Wraith, A. E. (1971). The mechanism of the bloomery process in shaft furnaces. *JISI*. 209. pg. 342-364.

Tylecote, Ronald F. and Cherry, John. (1970). The 17th century bloomery at Muncaster Head. *Transactions of the Cumberland and Westmorland Antiquarian Society.* 70. pg. 69-109.

Tylecote, R. F. and Clough, R. E. (1983). Recent bog-iron ore analysis and the smelting of pyrite nodules. *Offa*. 40. pg. 115-118.

Tylecote, R. F. and Merkel, J. F. (1985). Experimental smelting techniques: Achievements and future. In: *Furnaces and smelting technology in antiquity*. Craddock, P. T. and Hughes, M. J., eds. British Museum Occasional Paper 48, London, England. IS.

Tyler, P. M. (February 10, 1921). High speed steel manufacture in Sheffield. *Iron Age.* pg. 371-374.

United States Steel Corporation. (1957). *The making, shaping and treating of steel*. 7th edition. US. Steel Corp., Pittsburgh, PA.

Victoria and Albert Museum. (1913). *Old English pattern books of the metal trades.* Victoria and Albert Museum, London, England.

Villars, P. and Calvert, L. D. (1991). *Pearson's handbook of crystallographic data for intermetallic phases*. vols. 1-4. American Society for Metals, Metals Park, OH.

Wakelin, David H. ed. (1999). *The making, shaping, and treating of steel*. Vol. 1 Ironmaking volume. AISE Steel Foundation, Pittsburgh, PA.

Wanklyn, M. D. G. (1973). Iron and steelworks in Coalbrookdale 1645. *Shropshire Newsletter*. 44. pg. 3-6.

Wedel, Ernst von. (1960). The history of die forming. *Metal treatment and drop forging*. 27. pg. 401-408.

Wertime, Theodore A. (1962). *The coming of the age of steel*. Chicago, IL. IS.

Wertime, T. A. (1964). Man's first encounter with metallurgy. *Science*. 146. pg. 1257-1267. 159. pg. 927-935.

Wertime, Theodore A. and Muhly, James D., eds. (1980). *The coming of the age of iron.* Yale University Press, New Haven, CT. IS.

West, Thomas Dyson. (1902). *Metallurgy of cast iron: Complete exposition of the processes involved in its treatment, chemically and physically, from the blast furnace through the foundry to the testing machine*. Cleveland Print. and Pub. Co., Cleveland, OH.

Whitaker, P. and Williams, T. H. (1969). Examination of a north Indian sword in the collection of the Pitt-Rivers Museum, Oxford and experiments in the forging of low carbon strip. *Bull. HMG.* 3(2). pg. 39-45.

Williams, A. R. (1976). Ancient steel from Egypt. *Journal of Archaeological Science.* 3. pg. 294-301.

Wilson, A. J. C., ed. (1992). *International tables for crystallography.* Vol C. Kluwer Academic Publishers, Boston, MA.

Wolf, Fridolin. (2006). *The abcs of blacksmithing.* Blue Moon Press, Huntingdon, PA.

Woodworth, Joseph Vincent. (1903). *Hardening, tempering, annealing and forging of steel: A treatise on the practical treatment and working of high and low grade steel.* N.W. Henley & Co., NY, NY. IS.

Woodworth, Joseph Vincent. (1905). *American tool making & interchangeable manufacturing: A treatise upon the designing, constructing, use, and installation of tools, jigs, fixtures ... and labor-saving contrivances.* N.W. Henley & Co., NY, NY.

Woodworth, Joseph Vincent. (1911). *Drop forging, die sinking and machine forming of steel: Modern shop practices, processes, methods, machines, tools and details.* Munn & Co., NY, NY. IS.

Wright, J. (1907). *Electric furnaces and their industrial applications.* NY, NY.

Wynne, E. J. and Tylecote, R. F. (1956). An experimental investigation into primitive iron smelting technique. *Journal of the Iron and Steel Institute.*

Zastrow, Nancy B. (2004). *Lives shaped by steel: Celebrating east coast outdoor metal artists.* Iron Artists, Silver Springs, MD.

Zastrow, Nancy B. (2010). *Passion and power: Metal artists in western U.S.* Copper Heron, Silver Springs, MD.

Zapffe, Carl A. (Oct. 14, 1948). Who discovered stainless steel? *Iron Age.* CLXII.

www.ingramcontent.com/pod-product-compliance
Lightning Source LLC
Chambersburg PA
CBHW051210200326
41519CB00025B/7067